Aerospace Manufacturing Processes

Aerospace Manufacturing Processes

Editor

Khamisi Zaman

scitus academics

Aerospace Manufacturing Processes
Edited by **Khamisi Zaman**

ISBN: 978-1-68117-713-7
Library of Congress Control Number: 2016958813

SCITUS Academics LLC,
616, Corporate Way, Suite 2, 4766,
Valley Cottage, NY 10989

www.scitusacademics.com

Notice

Printed in United States of America on Acid Free Paper

Preface

The processes in which the raw materials are transformed into a final product are the steps. The process of production begins with the development of the materials from which the style is produced. These materials are then modified to form the necessary part through the production processes. The production processes can include treatment, machining or possibly reshaping of the material. The manufacturing process also includes quality assurance checks and tests during and after manufacturing and pre-production planning of the production process.

This book is aimed to expose students to modern aircraft manufacturing concepts and practices. Manufacturing operations for aircraft components include large activities involving several technologies for processing materials. It focuses on the presentation of production process technologies solely for the manufacture of major aircraft components. A wide range of aerospace materials coverage and processing are included with practical case studies and examples. The manufacturing process is undergoing major transformations due to the unexpected challenges arising from the current trend of miniaturization, the growth of new materials and the growing interaction between engineers and biologists in order to discover more about nature and living things. A "topdown" approach has traditionally been used in manufacturing. Engineers and scientists have recently begun to explore "bottom-up" approaches to the production of today's highly complicated products. In addition, these emerging processes are aimed at improving the quality and efficiency of the product. This book covers current and future directions for research, implementation and development of manufacturing processes in order to advance state-of-the-art manufacturing processes in aircraft industry and encourage innovation in the development of efficient and new processes.

Table of Contents

1

CHAPTER

A Review of Additive Manufacturing

Kaufui V. Wong and Aldo Hernandez

Department of Mechanical and Aerospace Engineering, University of Miami, Coral Gables, FL 33146, USA

ABSTRACT

Additive manufacturing processes take the information from a computer-aided design (CAD) file that is later converted to a stereolithography (STL) file. In this process, the drawing made in the CAD software is approximated by triangles and sliced containing the information of each layer that is going to be printed. There is a discussion of the relevant additive manufacturing processes and their applications. The aerospace industry employs them because of the possibility of manufacturing lighter structures to reduce weight. Additive manufacturing is transforming the practice of medicine and making work easier for architects. In 2004, the Society of Manufacturing Engineers did a classification of the various technologies and there are at least four additional significant technologies in 2012. Studies are reviewed which were about the strength of products made in additive manufacturing processes. However, there is still a lot of work and research to be accomplished before additive manufacturing technologies become standard in the manufacturing industry because not every commonly used manufacturing material can be handled. The accuracy needs improvement to eliminate the necessity of a finishing process. The continuous and increasing growth experienced since the early days and the successful results up to the present time allow for optimism that additive manufacturing has a significant place in the future of manufacturing.

1. RAPID PROTOTYPING

The first form of creating layer by layer a three-dimensional object using computer-aided design (CAD) was rapid prototyping, developed in the 1980's for creating models and prototype parts. This technology was created to help the realization of what engineers have in mind. Rapid prototyping is one of the earlier additive manufacturing (AM) processes. It allows for the creation of printed parts, not just models. Among the major advances that this process presented to product development are the time and cost reduction, human interaction, and consequently the product development cycle [1], also the possibility to create almost any shape that could be very difficult to machine. However, at the present time it is not yet adopted in the manufacturing sector, but scientists, medical doctors, students and professors, market researchers, and artists use it [2–4]. With rapid prototyping, scientists and students can rapidly build and analyze models for theoretical comprehension and studies. Doctors can build a model of a damaged body to analyze it and plan better the procedure, market researchers can see what people think of a particular new product, and rapid prototyping makes it easier for artists to explore their creativity.

The steps involved in product development using rapid prototyping are shown in Figure 1. Here, it can be seen that creating models faster save a lot of time and there is the possibility of testing more models.

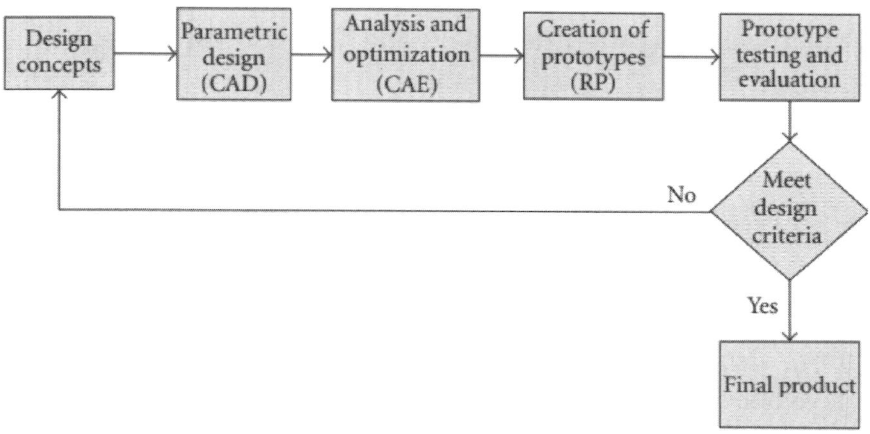

Figure 1: Product development cycle [2].

At the present time, the technologies of rapid prototyping are not just used for creating models, with the advantages in plastic materials it has been possible to create finished products, of course at the beginning they were developed to expand the situations tested in the prototyping process [5]. Nowadays, these technologies have other names like 3D printing, and so forth, but they all have the origins of rapid prototyping [2, 6]. According to Wohler's report 2011 the growth rate for 2010 was 24.1%. The compound annual growth rate for the industry's history, until 2010, is 26.2 percent [7, 8]. This growth shown in Figure 2.

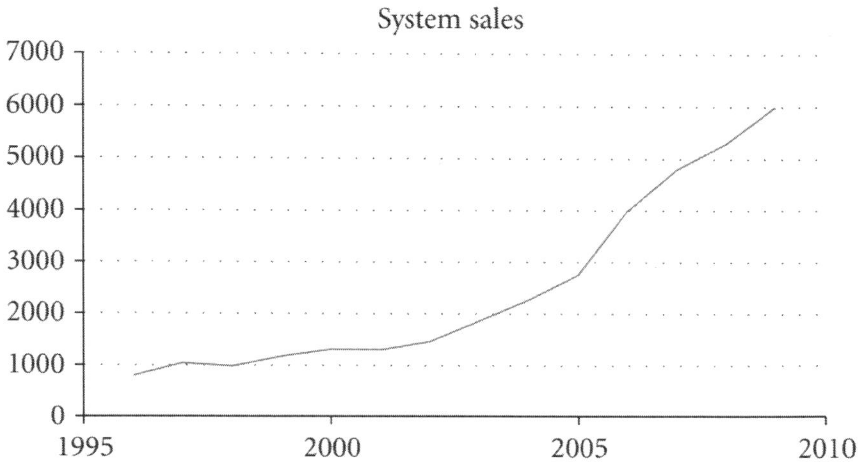

Figure 2: Growth of rapid prototyping [9]. Source. Wohler's report 2010 adapted from [10].

In addition, it is important to notice that rapid manufacturing became possible by other technologies, which are computer-aided design (CAD), computer-aided manufacturing (CAM), and computer numerical control (CNC). This three technologies combined together made possible the printing of three-dimensional objects [2,5, 11].

Rapid prototyping is still not the best solution for all cases, in some cases CNC machining processes still need to be used. Parts dimension could be larger than available additive manufacturing printers [8]. Materials for rapid prototyping are still limited. It is clear that at least it is possible to print metals and ceramics but not all commonly used manufacturing materials [11].

In Figure 3, there is an overview of the different additive manufacturing processes that are going to be further discussed. Here in this figure adapted from [11], the criterion used is to classify these processes into liquid base, solid based, and powder based. The processes included in this review are considered the most relevant in the past, and promising for the future of the industry. The processes considered are stereolithography (SL), Polyjet, fused deposition modeling (FDM), laminated object manufacturing (LOM), 3D printing (3DP), Prometal, selective laser sintering (SLS), laminated engineered net shaping (LENS), and electron beam melting (EBM). The liquid-and powder-based processes seem more promising than solid-based processes of which LOM is the predominant one today. In 2004 [11], EBM, Prometal, LENS, and Polyjet were nonexistent.

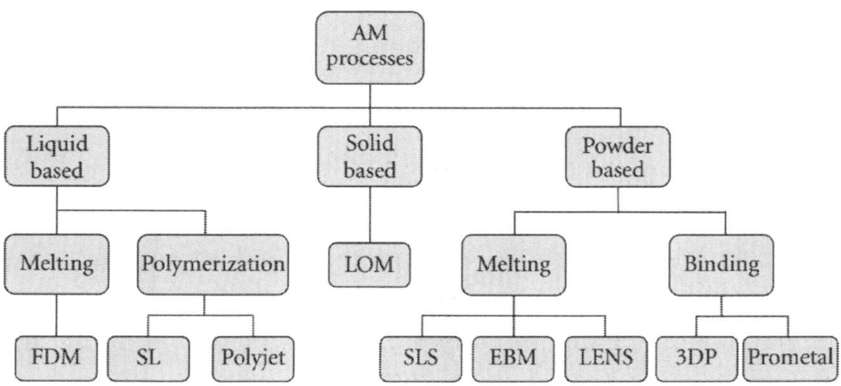

Figure 3: Three-dimensional printing processes. Adapted from [11].

These technologies were first created to produce models, but they have expanded since then. In the chart is presented a survey made by Wohlers in which 24 manufacturers participated and so did 65 services of 5000+ users and costumers. In Figure 4 is shown the amount of responses received by these companies [12].

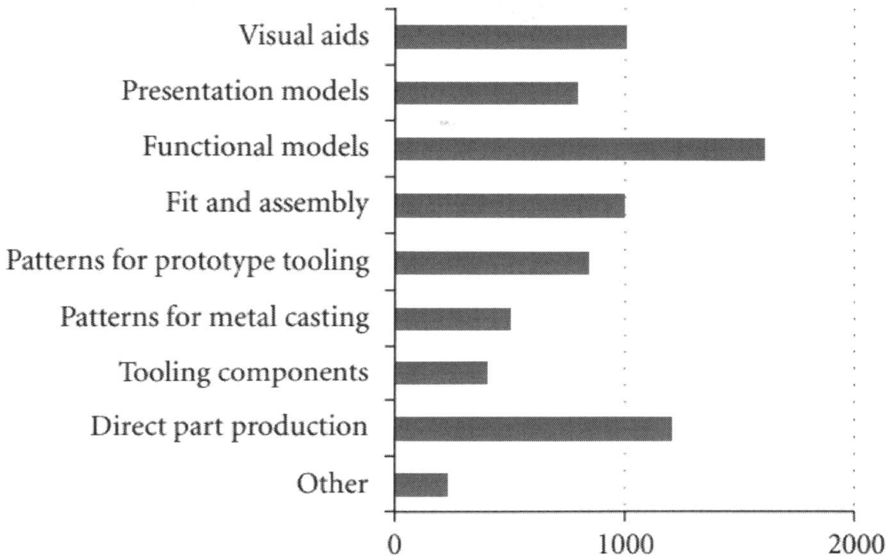

Figure 4: Different usage for additive manufacturing printing processes [12].

2. STEREOLITHOGRAPHY

Stereolithography (SL), developed by 3D Systems, Inc., was the first and is most widely used process of rapid prototyping, so, in the past the two terms were used synonymously. This is a liquid-based process that consists in the curing or solidification of a photosensitive polymer when an ultraviolet laser makes contact with the resin. The process starts with a model in a CAD software and then it is translated to a STL file in which the pieces are "cut in slices" containing the information for each layer. The thickness of each layer as well as the resolution depend on the equipment used. A platform is built to anchor the piece and supporting the overhanging structures. Then the UV laser is applied to the resin solidifying specific locations of each layer. When the layer is finished the platform is lowered and finally when the process is done the excess is drained and can be reused [2, 5, 11]. A newer version of this process has been developed with a higher resolution and is called microstereolithography. This process that has a layer thickness of less than 10 µm can be achieved [13]. In Figure 5 are shown the basic parts of a stereolithography machine.

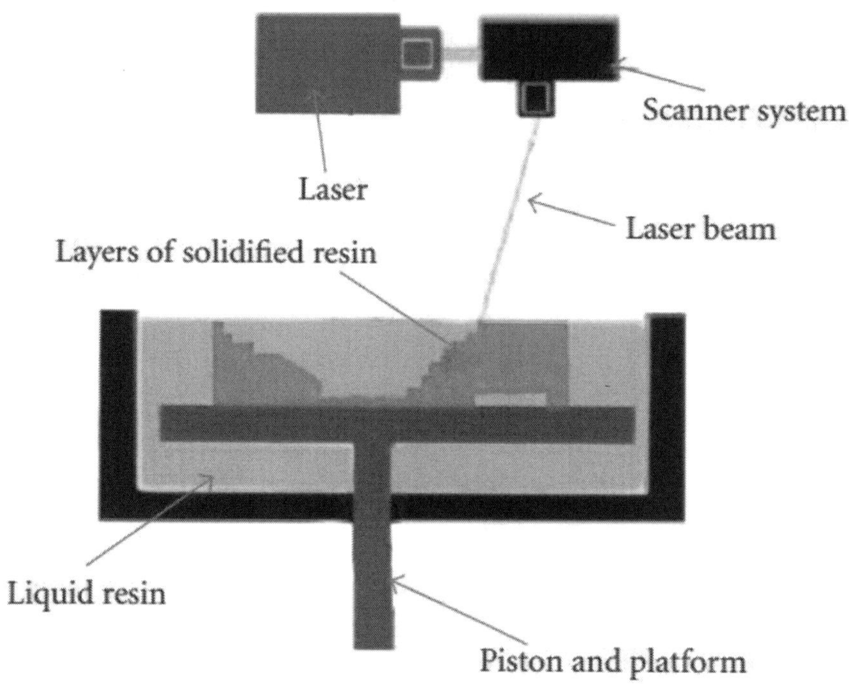

Figure 5: Stereolithography.

The basic principle of this process is the photopolymerization, which is the process where a liquid monomer or a polymer converts into a solidified polymer by applying ultraviolet light which acts as a catalyst for the reactions; this process is also called ultraviolet curing. It is also possible to have powders suspended in the liquid like ceramics [14].

There are errors induced to the final piece from the process of stereolithography. One is overcuring, which occurs to overhang parts because there is no fusing with a bottom layer. Another is the scanned line shape, which is introduced by the scanning process. Because the resin is a high-viscosity liquid the layer thickness is variable and this introduces an error in the border position control. Another error caused could be if the part needed to have a surface finished process that is normally done by hand [21]. All these errors are minimized in equipments of high quality.

There is the possibility of using different materials while building a piece; this process is called multiple material stereolithography. In order to print with different materials all the resin has to be drained and filled with the new material when the process reach the layer where the change is going to take place. This must occuer even if the first material is going to be used again because is only possible to print consecutive layers. resin. In the software a scheduling process has to be specified [22].

3. THE STL FILE

The STL file was created in 1987 by 3D Systems Inc. when they first developed the stereolithography, and the STL file stands for this term. It is also called Standard Tessellation Language. There are other types of files, but the STL file is the standard for every additive manufacturing process. The STL file creation process mainly converts the continuous geometry in the CAD file into a header, small triangles, or coordinates triplet list of $x, y,$ and z coordinates and the normal vector to the triangles. This process is inaccurate and the smaller the triangles the closer to reality [2, 13, 23]. The interior and exterior surfaces are identified using the right-hand rule and vertices cannot share a point with a line. Additional edges are added when the figure is sliced. The slicing process also introduces inaccuracy to the file because here the algorithm replaces the continuous contour with discrete stair steps [23]. To reduce this inaccuracy, the technique for a feature that has a small radius in relation to the dimension of the part is to create STL files separately and to combine them later. The dimension in *z* direction should be designed to have a multiple of the layer thickness value [21]. In Figure 6 is shown the position of the STL file creation in the data flow of a rapid prototyping process. In Figure 7 is shown the data flow in the STL file creation software.

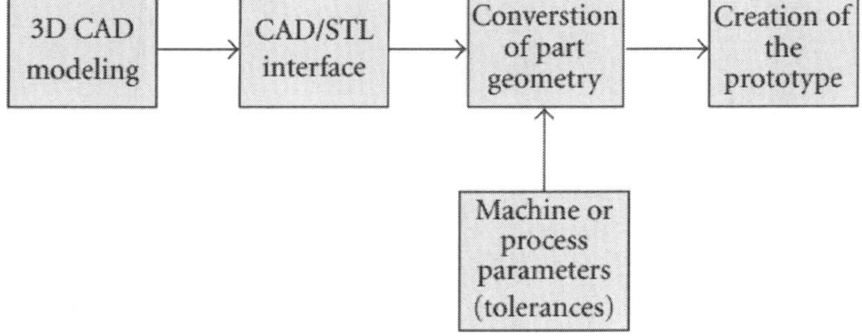

Figure 6: Data flow in rapid prototyping [2].

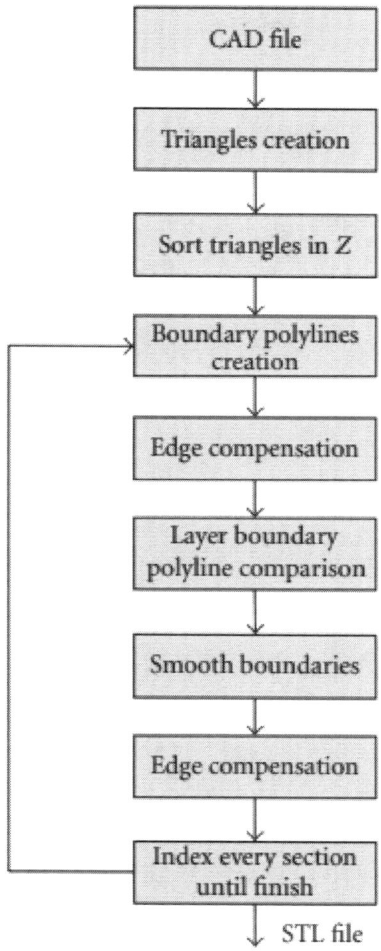

Figure 7: Data flow in STL file creation [2].

Other types of files are stereolithography contour (SLC) and SLI from 3D Systems, CLI from EOS, Hewlett-Packard graphics language (HPGL) from Hewlett-Packard, stereolithography contour from Stratasys, and F&S from Fockele and Schwarze and initial graphics exchange specifications (IGES) [13].

4. 3DP

3DP process is a MIT-licensed process in which water-based liquid binder is supplied in a jet onto a starch-based powder to print the data from a CAD drawing. The powder particles lie in a powder bed and they are glued together when the binder is jetted. This process is called 3DP because of the similarity with the inkjet printing process that is used for two-dimensional printing in paper. This process can handle a high variety of polymers [5, 13].

5. FUSED DEPOSITION MODELING

Fused deposition modeling (FDM) is an additive manufacturing process in which a thin filament of plastic feeds a machine where a print head melts it and extrude it in a thickness typically of 0.25 mm. Materials used in this process are polycarbonate (PC), acrylonitrile butadiene styrene (ABS), polyphenylsulfone (PPSF), PC-ABS blends, and PC-ISO, which is a medical grade PC. The main advantages of this process are that no chemical post-processing required, no resins to cure, less expensive machine, and materials resulting in a more cost effective process [2, 5]. The disadvantages are that the resolution on the *z* axis is low compared to other additive manufacturing process (0.25 mm), so if a smooth surface is needed a finishing process is required and it is a slow process sometimes taking days to build large complex parts. To save time some models permit two modes; a fully dense mode and a sparse mode that save time but obviously reducing the mechanical properties [24]. In Figure 8 is shown the basics fused deposition modeling process.

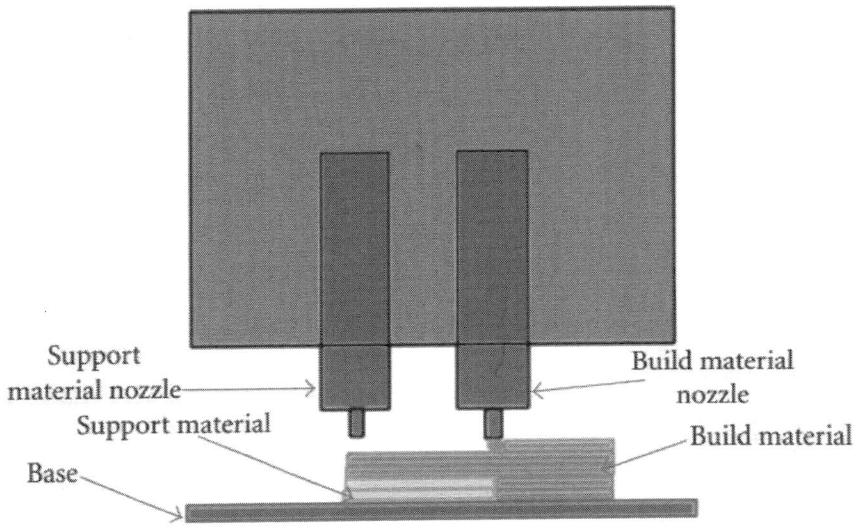

Figure 8: Fused deposition modeling.

6. PROMETAL

Prometal is a three-dimensional printing process to build injection tools and dies. This is a powder-based process in which stainless steel is used. The printing process occurs when a liquid binder is spurt out in jets to steel powder. The powder is located in a powder bed that is controlled by build pistons that lowers the bed when each layer is finished and a feed piston that supply the material for each layer. After finishing, the residual powder must be removed. When building a mold no postprocessing is required. If a functional part is being built, sintering, infiltration, and finishing processes are required [5, 11]. In the sintering process, the part is heated to 350°F for 24 hour hardening the binder fusing with the steel in a 60% porous specimen. In the infiltration process, the piece is infused with bronze powder when they are heated together to more than 2000°F in an alloy of 60% stainless steel and 40% bronze [25–27]. The same process, but with different sintering temperatures and times, has been used with other materials like a tungsten carbide powder sintered with a zirconium copper alloy for the manufacturing of rocket nozzles; these parts have better properties than CNC machined parts of the same material [28].

7. SELECTIVE LASER SINTERING

This is a three-dimensional printing process in which a powder is sintered or fuses by the application of a carbon dioxide laser beam. The chamber is heated to almost the melting point of the material. The laser fused the powder at a specific location for each layer specified by the design. The particles lie loosely in a bed, which is controlled by a piston, that is lowered the same amount of the layer thickness each time a layer is finished. This process offers a great variety of materials that could be used: plastics, metals, combination of metals, combinations of metals and polymers, and combinations of metals and ceramics [13, 29, 30]. Examples of the polymers that could be used are acrylic styrene and polyamide (nylon), which show almost the same mechanical properties as the injected parts [29, 31]. It is also possible to use composites or reinforced polymers, that is, polyamide with fiberglass. They also could be reinforced with metals like copper. For metals, a binder is necessary. This could be a polymer binder, which will be later removed by heating or a mix of metals with very different melting point [29, 31, 32]. Parts of alumina with high strength can be built with polyvinyl alcohol, which is an organic binder [5]. The main advantages of this technology are the wide range of materials that can be used. Unused powder can be recycled. The disadvantages are that the accuracy is limited by the size of particles of the material, oxidation needs to be avoided by executing the process in an inert gas atmosphere and for the process to occur at constant temperature near the melting point. This process is also called direct metal laser sintering.

8. ELECTRON BEAM MELTING

A process similar to SLS is electron beam melting (EBM). This process is relatively new but is growing rapidly. In this process, what melts the powder is an electron laser beam powered by a high voltage, typically 30 to 60 KV. The process takes place in a high vacuum chamber to avoid oxidation issues because it is intended for building metal parts. Other than this, the process is very similar to SLS. EBM also can process a high variety of prealloyed metals. One of the future uses of this process is the manufacturing in outer space [33, 34], since it is all done in a high vacuum chamber.

9. LASER ENGINEERED NET SHAPING

In this additive manufacturing process, a part is built by melting metal powder that is injected into a specific location. It becomes molten with the use of a high-powered laser beam. The material solidifies when it is cooled down. The process occurs in a closed chamber with an argon atmosphere. This process permits the use of a high variety of metals and combination of them like stainless steel, nickel-based alloys, titanium-6 aluminium-4 vanadium, tooling steel, copper alloys, and so forth. Alumina can be used too. This process is also used to repair parts that by other processes will be impossible or more expensive to do. One problem in this process could be the residual stresses by uneven heating and cooling processes that can be significant in high-precision processes like turbine blades repair [5, 13, 35–37]. Figure 9 is an illustration of how the part is made in this process.

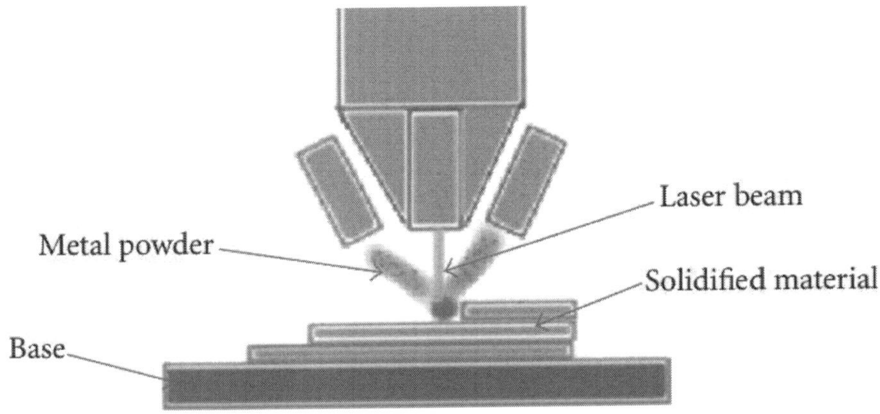

Figure 9: Laser engineered net shaping.

10. LAMINATED OBJECT MANUFACTURING

Laminated Object Manufacturing (LOM) is a process that combines additive and subtractive techniques to build a part layer by layer. In this process the materials come in sheet form. The layers are bonded together by pressure and heat application and using a thermal adhesive coating. A carbon dioxide laser cuts the material to the shape of each layer given the information of the 3D model from the CAD and STL file. The advantages of this process are the low cost, no post processing and supporting structures required, no deformation or phase change during the process, and the possibility of building large parts. The disadvantages are that the fabrication material is subtracted thus wasting it, low surface definition, the material is directional dependent for machinability and mechanical properties, and complex internal cavities are very difficult to be built. This process can be used for models with papers, composites, and metals [2, 5, 38].

11. POLYJET

This is an additive manufacturing process that uses inkjet technologies to manufacture physical models. The inkjet head moves in the x and y axes depositing a photopolymer which is cured by ultraviolet lamps after each layer is finished. The layer thickness achieved in this process is 16 μm, so the produced parts have a high resolution. However, the parts produced by this process are weaker than others like stereolithography and selective laser sintering. A gel-type polymer is used for supporting the overhang features and after the process is finished this material is water jetted. With this process, parts of multiple colors can be built [9, 39, 40].

12. APPLICATIONS

12.1. Lightweight Machines

With additive manufacturing technologies it is possible to manufacture lightweight parts. In the automotive and aerospace industry the main goal is to make the lightest practical car or aircraft while securing safety. Additive manufacturing technologies have enabled the manufacture of complex cross sectional areas like the honeycomb cell [41] or every other material part that contains cavities and cut-outs which reduce the weight-strength relation. It is possible to create lightweight structures; they are methods to get a shape that have a minimum weight like the hanging method and the soap film method [42]. The hanging method and the soap fill method produce a very difficult form of a structure which has been used for civil construction, but with additive manufacturing it is possible to create structural parts for machines using the shape described by these methods and reducing the total weight. Selective laser sintering, and electron beam are now used in the aircraft and aerospace industries. Engineers perform design within the manufacturing constrains but this process expands the limits. With SLS and EBM, the limit will be the engineer's imagination. They open a whole new dimension of possible designs with almost any prealloyed metal powder [37]. With the traditional process these complex shape structures will be expensive to do if at all possible. With additive manufacturing printing technologies like selective laser sintering or electron beam melting, hollow structures, which are less expensive than a solid one, can be made since less material is used.

12.2. Architectural Modeling

Creating an architectural model can be very difficult for architects. Architects usually build their models with hand techniques, but when complex models are on their minds making a physical model can be a very hard task. Modeling is very important for the architects to study the models and their functionality. They are also needed for architects to explain them to their customers and convince them to make the project a reality [43]. Additive manufacturing technologies can provide architects a very powerful tool for their business, by being able to create a physical model faster without worrying about the complexity of their design. It also achieves a better resolution than other processes used in architecture. Architects work with CAD software, so there is no need for them to adapt to anything because the STL file is created from a CAD file. Stereolithography is a process very suitable for the architectural modeling because of the materials used and the printing resolution [34–47].

12.3. Medical Applications

Additive manufacturing printing technologies have vast applications in the medical world. They are transforming the practice of medicine through the possibilities of making rapid prototypes and very high quality bone transplants and models of damaged bone of the patients for analysis. Additive manufacturing printing methods permit to scan and build a physical model of defective bones from patients and give doctors a better idea of what to expect and plan better the procedure, this will save cost and time and help achieve a better result [48, 49]. Bone transplants now can be done by printing them and additive manufacturing methods make it possible to have a transplant that is practically identical to the original. Because of the limitless form or shape of what could be built, doctors have the option to create a porous-controlled material that will permit osteoconductivity or to create a precise metal transplant identical to the original depending on the bone to be replaced [50–52]. Characteristics of the transplants such as density, pore shape and size, and pore interconnectivity are important parameters that will manipulate tissue ingrowth and mechanical properties of the implant bone. The mechanical strength of these implants are three to five times higher than others produced by other processes and the possibility of inflammation caused by microdebris that breaks during the procedure is reduced [45]. Additive manufacturing is a very good tool for dentists because they can easily build a plaster model of a patient's mouth or replace the teeth, which have a unique form with process like stereolithography, selective laser sintering and electron beam melting [53–55].

According to PC Magazine [54], an 83-year-old Belgian woman became the first-ever person to receive a transplant jawbone tailor-made for her face using a 3D printer and the surgery time and recovery were a lot less than other patients that received the same procedure. The shapes of bones differ too much between each person and additive manufacturing printing produces transplants that fit better, and are easier to insert and secure, reducing the time for the procedure and produce a better cosmetic result [55, 56].

Stereolithography is being used to manufacture prosthetic sockets. By using this technology to ensure that the form of the socket adapt better to the patient while being more cost-effective than hand or machined methods [57]. Not only hard parts like bones can be produced, also it is possible to print cells in a 3D array that with the possibility of printing complex shapes and arrays human tissue can be printed [58–61]. This technology will help patients that have lost tissue in accidents or from other reasons to recover faster and with better cosmetic results. In addition 3D cell printing technologies offer the possibility of printing artificial blood vessels that can be used in the coronary bypass surgery or any other blood vessel procedure or diseases, like cardiovascular defects and medical therapy [62–64]. The application of this printed blood vessels is in the future. Research in this area, also called bioprinting organs, will eventually lead to printed organs, but this could take 20 years until someone achieves it [65].

Cell printing is not limited to print human tissue; it is also used in the field of molecular electronics. The precision of high-resolution processes like nanolithography and photolithography permits the creation of biochips and biosensors [63, 66, 67].

12.4. Improving the Manufacturing of Fuel Cells

Additive manufacturing technologies can be used in processes that require a very precise thin film of a certain material. In the manufacture of polymer electrolyte membrane fuel cells (PEMFCs), it is necessary to precisely deposit a very thin layer of platinum, needed for the oxidation and reduction reactions, with high utilization efficiency of the platinum. This could be critical for taking this technology to the masses. One of the other processes used is screen printing, but this process is done by hand and compromises the uniformity and time efficiency. The process is greatly improved by using the 3D printing process to deposit the layer of platinum. In Figure 10, there is a comparison between the processes. The inkjet printing method is 4 to 5 times faster than screen printing [15].

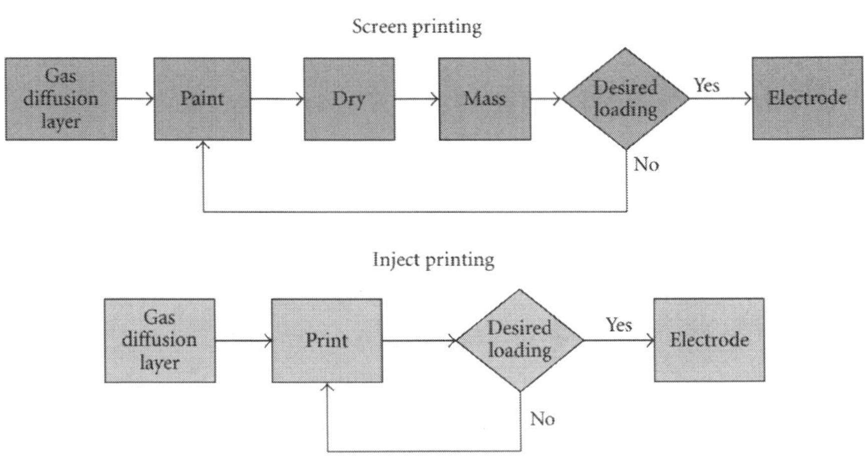

Figure 10: Screen printing compared to inkjet Printing [15].

12.5. Additive Manufacturing in Art

Additive manufacturing technologies are a very powerful tool to artist in the fashion, furniture, and lightning industry given the possibility of virtually manufacturing the most complex form imaginable. There are companies that manufacture furnishing complements, lightning, and accessories including clothes using SLS [68–70].

12.6. Additive Manufacturing for Hobbyist

Additive manufacturing technologies are reaching nonindustrial users. This revolution started in 2007 with printers that could be as low as $500 using a variant of the fused deposition modeling process. However, these low-cost printers are mostly sold as do-it-yourself (DIY) projects, so, technical ability on the part of the users is required. However, these days with larger companies entering the business 3D printers are closer to reaching the masses. At this time, plastics and edible materials like chocolate materials are only available [71–73]. Users can use CAD software normally used by engineers. Easier ones for hobbyists are available which can be used for design. There are many apps that hobbyists can make 3D models and print them using a 3D printer [74].

12.7. Strength Comparison

Studies have been made to analyze the properties of the product in each process. Kim and Oh [16] compared the properties of nonmetal additive manufacturing processes. They tested the specimens in the building direction and perpendicular to the building direction and they found very little influence in the building direction in 3DP but an enormous influence in LOM [70].

Figure 11 is a comparison of the strength for LOM, Polyjet, SL, SLS, FDM, and 3DP.

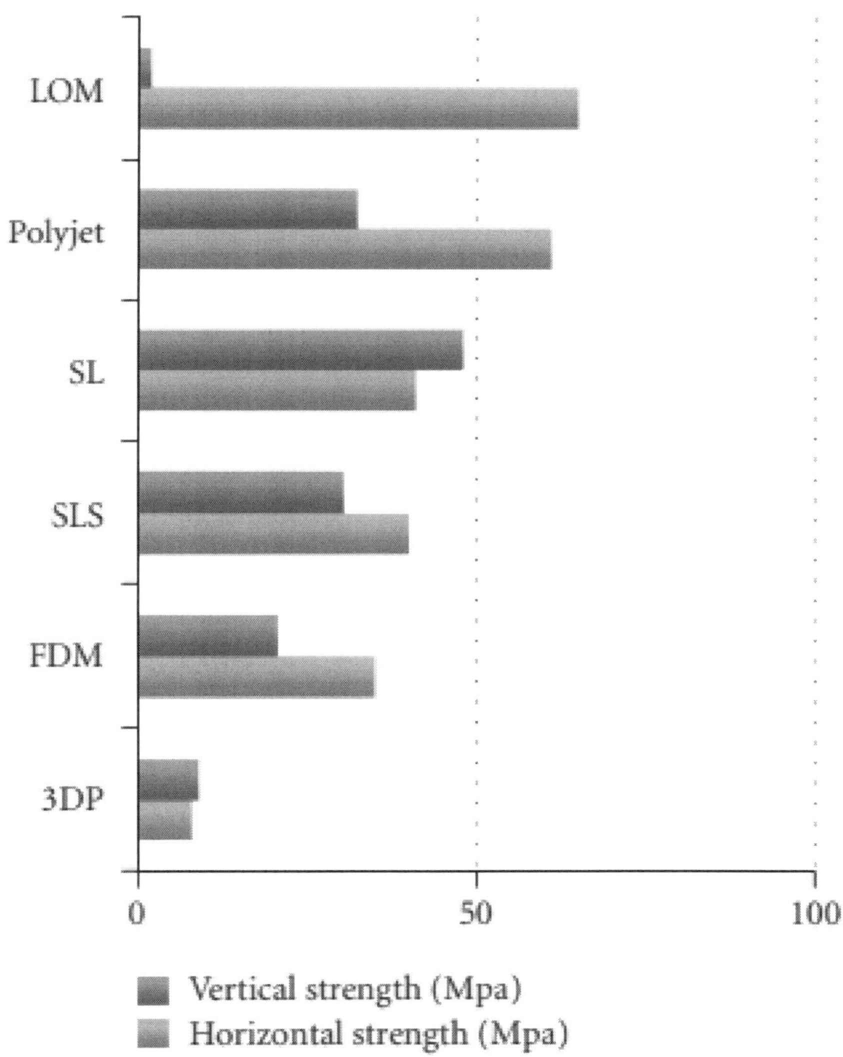

Figure 11: Tensile strength of various 3D printing processes adapted from [16].

The strength of the material produced by Prometal, EBM, SLS, and LENS is compared in Figure 12. The studies were made by [17–20].

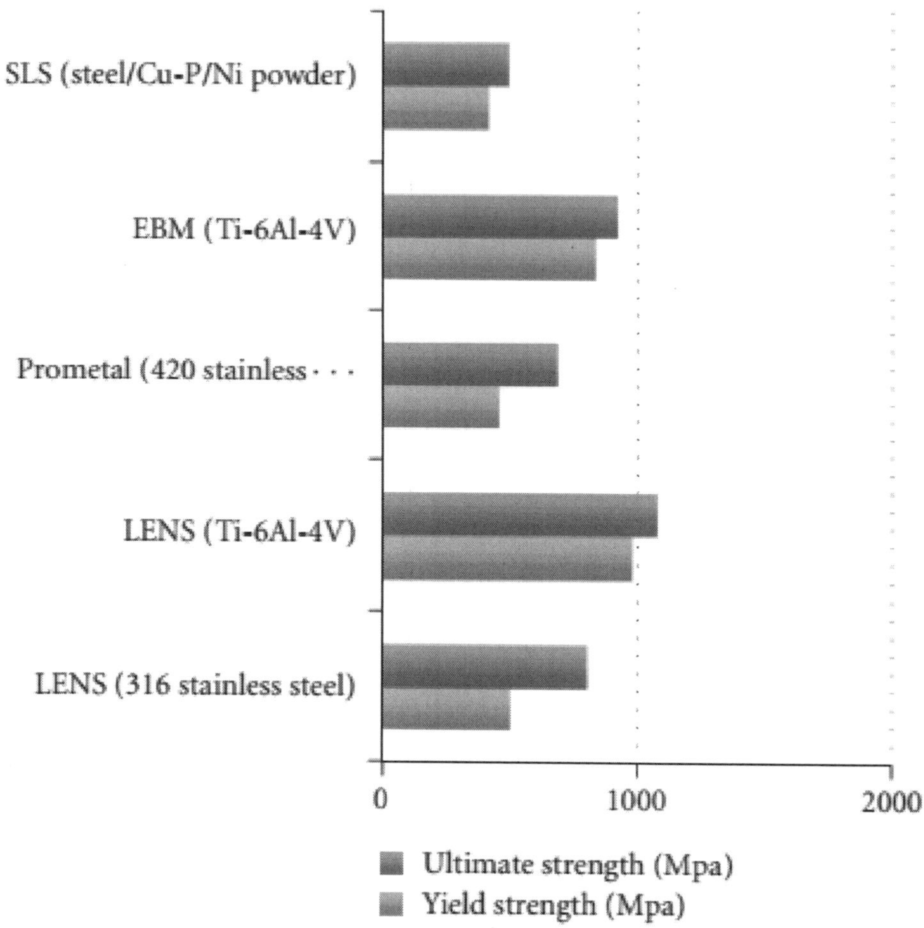

Figure 12: Tensile strength of various 3D printing processes [17–20].

13. DISCUSSION AND CONCLUSION

In this paper article is discussed the early versions of additive manufacturing for making fast prototypes that was initiated by the necessity of speeding the process in model development and shortening the time between product development and market placement. Additive manufacturing processes take the information from a CAD file that is later converted to an STL file. In this process, the drawing made in the CAD software is approximated by triangles and sliced containing the information of each layer that is going to be printed. There is also a discussion of the relevant additive manufacturing processes and their applications and a review of how the parts are made using these additive manufacturing processes. The continuous and increasing growth experienced since the early days and the successful results up to date, there is optimism that additive manufacturing has a significant place in the future of manufacturing.

In 2004, the Society of Manufacturing Engineers did a classification of the various technologies [11], but there are at least four additional significant technologies in 2012.

Additive manufacturing technologies have been welcomed in the aerospace industry because of the possibility to manufacture lighter structures to reduce weight, which is the common goal of aircraft and spacecraft designers. In the automotive industry, additive manufacturing is advantageous also in reproducing difficult-to-find parts, for example, parts for classic cars. Additive manufacturing is transforming the practice of medicine; now it is possible to have a precise model of a bone before a surgery and the possibility of creating an accurate transplant, no matter how complex its form is. Additive manufacturing is making work easier for architects, who now can print the 3D models of whatever complex shape for a civil project they have in mind. In addition, studies are reviewed which were about the strength of products made in additive manufacturing processes. Review has been presented of studies considering nonmetal material processes and metal material processes in which a comparison of strength of the products was made by the various processes. However, there is still a lot of work and research to be accomplished before additive manufacturing processes become the standard in the manufacturing industry because not every commonly used manufacturing material can be handled. The accuracy needs improvement to eliminate the necessity of a finishing process and to be able to produce parts that require the highest levels of precision.

REFERENCES

1. S. Ashley, "Rapid prototyping systems," Mechanical Engineering, vol. 113, no. 4, p. 34, 1991.

2. R. Noorani, Rapid Prototyping—Principles and Applications, John Wiley & Sons, 2006.

3. J. Flowers and M. Moniz, "Rapid prototyping in technology education," Technology Teacher, vol. 62, no. 3, p. 7, 2002.

4. C. K. Chua, S. M. Chou, S. C. Lin, K. H. Eu, and K. F. Lew, "Rapid prototyping assisted surgery planning,"International Journal of Advanced Manufacturing Technology, vol. 14, no. 9, pp. 624–630, 1998.

5. K. Cooper, Rapid Prototyping Technology, Marcel Dekker, 2001.

6. A. Kochan, "Rapid growth for rapid prototyping," Assembly Automation, vol. 17, no. 3, pp. 215–217, 1997.

7. T. Wohlers, Wohlers Report 2011, Wholers Associates, 2011.

8. T. Wohlers, "Additive Manufacturing Advances," Manufacturing Engineering, vol. 148, no. 4, pp. 55–56, 2012.

9. T. Wohlers, Wohlers Report 2010, Wholers Associates, 2010.

10. T. Grimm, User's Guide to Rapid Prototyping, Society of Manufacturing Engineers, 2004.

11. P. P. Kruth, "Material incress manufacturing by rapid prototyping techniques," CIRP Annals—Manufacturing Technology, vol. 40, no. 2, pp. 603–614, 1991.

12. T. Wohlers, Wohlers Report 2009, Wholers Associates, 2009.

13. J. W. Halloran, V. Tomeckova, S. Gentry et al., "Photopolymerization of powder suspensions for shaping ceramics," Journal of the European Ceramic Society, vol. 31, no. 14, pp. 2613–2619, 2011.

14. D. T. Pham and C. Ji, "Design for stereolithography," Proceedings of the Institution of Mechanical Engineers, vol. 214, no. 5, pp. 635–640, 2000.

15. A. D. Taylor, E. Y. Kim, V. P. Humes, J. Kizuka, and L. T. Thompson, "Inkjet printing of carbon supported platinum

3-D catalyst layers for use in fuel cells," Journal of Power Sources, vol. 171, no. 1, pp. 101–106, 2007.

16. G. D. Kim and Y. T. Oh, "A benchmark study on rapid prototyping processes and machines: quantitative comparisons of mechanical properties, accuracy, roughness, speed, and material cost," Proceedings of the Institution of Mechanical Engineers, vol. 222, no. 2, pp. 201–215, 2008.

17. J. P. Kruth, X. Wang, T. Laoui, and L. Froyen, "Lasers and materials in selective laser sintering," Assembly Automation, vol. 23, no. 4, pp. 357–371, 2003.

18. L. Facchini, E. Magalini, P. Robotti, and A. Molinari, "Microstructure and mechanical properties of Ti-6Al-4V produced by electron beam melting of pre-alloyed powders," Rapid Prototyping Journal, vol. 15, no. 3, pp. 171–178, 2009.

19. R. Shivpuri, X. Cheng, K. Agarwal, and S. Babu, "Evaluation of 3D printing for dies in low volume forging of 7075 aluminum helicopter parts," Rapid Prototyping Journal, vol. 11, no. 5, pp. 272–277, 2005.

20. Y. Xiong, Investigation of the laser engineered net shaping process for nanostructured cermets [ProQuest Dissertations], University of California, 2009.

21. H. Kim, C. Jae-Won, and R. Wicker, "Scheduling and process planning for multiple material stereolithography," Rapid Prototyping Journal, vol. 16, no. 4, pp. 232–240, 2010.

22. M. Szilvœi-Nagy and G. Mátyási, "Analysis of STL files," Mathematical and Computer Modelling, vol. 38, no. 7–9, pp. 945–960, 2003.

23. C. Iancu, D. Iancu, and A. Stamcioiu, "From Cad model to 3D print via "STL" file format," http://www.utgjiu.ro/rev_mec/mecanica/pdf/2010-01/13_Catalin%20Iancu.pdf.

24. S. Morvan, R. Hochsmann, and M. Sakamoto, "ProMetal RCT(TM) process for fabrication of complex sand molds and sand cores," Rapid Prototyping, vol. 11, no. 2, pp. 1–7, 2005.

25. R. C. T. ProMetal, "ProMetal RCT rapid prototyping and digital sand casting services," 2010, http://www.youtube.com/watch?v=Z8MaVaqNr3U.

26. Ex One, "3D metal printing," 2010, http://www.youtube.com/watch?v=i6Px6RSL9Ac&feature=related.

27. D. W. Lipke, Y. Zhang, Y. Liu, B. C. Church, and K. H. Sandhage, "Near net-shape/net-dimension ZrC/W-based composites with complex geometries via rapid prototyping and displacive compensation of porosity," Journal of the European Ceramic Society, vol. 30, no. 11, pp. 2265–2277, 2010.

28. J. P. Kruth, P. Mercelis, J. van Vaerenbergh, L. Froyen, and M. Rombouts, "Binding mechanisms in selective laser sintering and selective laser melting," Rapid Prototyping Journal, vol. 11, no. 1, pp. 26–36, 2005.

29. T. Hwa-Hsing, C. Ming-Lu, and Y. Hsiao-Chuan, "Slurry-based selective laser sintering of polymer-coated ceramic powders to fabricate high strength alumina parts," Journal of the European Ceramic Society, vol. 31, no. 8, pp. 1383–1388, 2011.

30. G. V. Salmoria, R. A. Paggi, A. Lago, and V. E. Beal, "Microstructural and mechanical characterization of PA12/MWCNTs nanocomposite manufactured by selective laser sintering," Polymer Testing, vol. 30, no. 6, pp. 611–615, 2011.

31. D. Slavko and K. Matic, "Selective laser sintering of composite materials technologies," Annals of DAAAM & Proceedings, p. p1527, 2010.

32. Technology Gateway, "NASA| EBF3—electron beam form fabrication," 2009, http://www.youtube.com/watch?v=WrWHwHuWrzk.

33. L. Murr, S. Gaytan, D. Ramirez et al., "Metal fabrication by additive manufacturing using laser and electron beam melting technologies," Journal of Materials Science & Technology, vol. 28, no. 1, pp. 1–14, 2012.

34. C. Semetay, Laser engineered net shaping (LENS) modeling using welding simulation concepts [ProQuest Dissertations and Theses], Lehigh University, 2007.

35. Y. Xiong, Investigation of the laser engineered net shaping process for nanostructured cermets [ProQuest Dissertations and Theses], University of California, 2009.

36. V. K. Balla, S. Bose, and A. Bandyopadhyay, "Processing of bulk alumina ceramics using laser engineered net shaping," International Journal of Applied Ceramic Technology, vol. 5, no. 3, pp. 234–242, 2008.

37. Y. S. Liao, H. C. Li, and Y. Y. Chiu, "Study of laminated object manufacturing with separately applied heating and pressing," International Journal of Advanced Manufacturing Technology, vol. 27, no. 7-8, pp. 703–707, 2006.

38. B. Vaupotic, M. Brezocnik, and J. Balic, "Use of PolyJet technology in manufacture of new product,"Journal of Achievements in Materials and Manufacturing Engineering, vol. 18, no. 1-2, pp. 319–322, 2006.

39. R. Singh, "Process capability study of polyjet printing for plastic components," Journal of Mechanical Science and Technology, vol. 25, no. 4, pp. 1011–1015, 2011.

40. V. Petrovic, J. Vicente, H. Gonzalez et al., "Additive layered manufacturing: sectors of industrial application shown through case studies," International Journal of Production Research, vol. 49, no. 4, pp. 1061–1079, 2011.

41. K. U. Bletzinger and E. Ramm, "Structural optimization and form finding of light weight structures,"Computers and Structures, vol. 79, no. 22–25, pp. 2053–2062, 2001.

42. A. Williams, "Architectural modelling as a form of research," Architectural Research Quarterly, vol. 6, no. 4, pp. 337–347, 2002.

43. SweetOnionsCreations, "Architecture model and 3D printing—sweet onion creations," 2007, http://www.youtube.com/watch?v=rEzugxybKmA.

44. M. Phair, "Rapid prototyping: the next wave in architectural modeling," Building Design & Construction, vol. 45, no. 11, pp. 15–16, 2004.

45. I. Gibson, T. Kvan, and W. Ling, "Rapid prototyping for architectural models," Rapid Prototyping Journal, vol. 8, no. 2, pp. 91–99, 2002.

46. J. Giannatsis, V. Dedoussis, and D. Karalekas, "Architectural scale modelling using stereolithography,"Rapid Prototyping Journal, vol. 8, no. 3, pp. 200–207, 2002.

47. F. Rengier, A. Mehndiratta, H. von Tengg-Kobligk et al., "3D printing based on imaging data: review of medical applications," International Journal of Computer Assisted Radiology and Surgery, vol. 5, no. 4, pp. 335–341, 2010.

48. W. J. James, M. A. Slabbekoorn, W. A. Edgin, and C. K. Hardin, "Correction of congenital malar hypoplasia using stereolithography for presurgical planning," Journal of Oral and Maxillofacial Surgery, vol. 56, no. 4, pp. 512–517, 1998.

49. Chaput, Christophe, and J. B. Lafon, "Ceramic industry," vol. 161, no. 9, pp. 15–16, 2011.

50. G. Fielding, A. Bandyopadhyay, and B. Susmita, "Effects of silica and zinc oxide doping on mechanical and biological properties of 3D printed tricalcium phosphate tissue engineering scaffolds," Dental Materials, vol. 28, no. 2, pp. 113–122, 2012.

51. J. Suwanprateeb, R. Sanngam, W. Suvannapruk, and T. Panyathanmaporn, "Mechanical and in vitro performance of apatite-wollastonite glass ceramic reinforced hydroxyapatite composite fabricated by 3D-printing," Journal of Materials Science, vol. 20, no. 6, pp. 1281–1289, 2009.

52. R. Makovec, "Digital technologies in dental laboratories," Annals of DAAAM & Proceedings, p. p1579, 2010.

53. R. van Noort, "The future of dental devices is digital," Dental Materials, vol. 28, no. 1, pp. 3–12, 2012.

54. A. Moscaritolo, "Woman receives 3D printer? created transplant Jaw," PC Magazine Online, 2012, http://www.pcmag.com/article2/0,2817,2399887,00.asp.

55. S. J. Hollister, "Porous scaffold design for tissue engineering," Nature Materials, vol. 4, no. 7, pp. 518–524, 2005.

56. M. A. Stoodley, J. R. Abbott, and D. A. Simpson, "Titanium cranioplasty using 3-D computer modelling of skull defects," Journal of Clinical Neuroscience, vol. 3, no. 2, pp. 149–155, 1996.

57. N. Herbert, D. Simpson, W. D. Spence, and W. Ion, "A preliminary investigation into the development of 3-D printing of prosthetic sockets," Journal of Rehabilitation Research and Development, vol. 42, no. 2, pp. 141–146, 2005.

58. B. Christensen, "New device prints human tissue," 2009, http://www.livescience.com/5977-device-prints-human-tissue.html.

59. T. Qian and Y. Wang, "Micro/nano-fabrication technologies for cell biology," Medical and Biological Engineering and Computing, vol. 48, no. 10, pp. 1023–1032, 2010.

60. S. J. Song, J. Choi, Y. D. Park et al., "Sodium alginate hydrogel-based bioprinting using a novel multinozzle bioprinting system," Artificial Organs, vol. 35, no. 11, pp. 1132–1136, 2011.

61. V. Mironov, N. Reis, and B. Derby, "Bioprinting: a beginning," Tissue Engineering, vol. 12, no. 4, pp. 631–634, 2006.

62. M. Conner, "3-D medical printer to print body parts," EDN, vol. 55, no. 3, p. 9, 2010.

63. L. Zhao, V. Lee, S. Yoo, G. Dai, and X. Intes, "The integration of 3-D cell printing and mesoscopic fluorescence molecular tomography of vascular constructs within thick hydrogel scaffolds," Biomaterials, vol. 33, no. 21, pp. 5325–5332, 2012.

64. "A step forward for artificial blood vessels," Expert Review of Cardiovascular Therapy 5.5, 817+, Academic OneFile, 2007.

65. J. Thilmany, "Printed life: the 3-D printing of living organs for transplant isn't far-fetched, it's almost here," Mechanical Engineering-CIME, 44+, Academic OneFile, 2012.

66. M. R. Dusseiller, D. Schlaepfer, M. Koch, R. Kroschewski, and M. Textor, "An inverted microcontact printing method on topographically structured polystyrene chips for arrayed micro-3-D culturing of single cells," Biomaterials, vol. 26, no. 29, pp. 5917–5925, 2005.

67. R. D. Piner, J. Zhu, F. Xu, S. Hong, and C. A. Mirkin, "'Dip-pen' nanolithography," Science, vol. 283, no. 5402, pp. 661–663, 1999.

68. T. Wohlers, "Making products by using additive manufacturing," Manufacturing Engineering, vol. 146, no. 4, pp. 70–74, 2011.

69. J. Malik, "Are 3D-printed fabrics the future of sustainable textiles?" in Ecouterre, 2010, http://www.ecouterre.com/are?3d?printed?fabrics ?the?future?of?sustainabletextiles/.

70. F. Tortul, "3D printed shoes awarded most creative at Mittelmoda," in Freedom of Creation, 2010, http://www.freedomofcreation.com/for/3d-printed-shoes-awarded-most-creative-at-mittelmoda.

71. P. Wallich, "3-D Printers Proliferate [Hands On]," IEEE Spectrum, vol. 47, no. 9, p. 23, 2010.

72. T. Bradshaw, "The printer that is transforming the world of creation," The Bath Chronicle, p. 45, 2011.

73. M. Simkins, "3-D printing a goo goo: conceive and build your own tasty designs with sticky food or clay,"Technology and Leanrning, p. 48, 2008.

74. Singh and Sandeep, Beginning Google SketchUp for 3D Printing, Springer, 2010.

2

CHAPTER

Rotorcraft Design for Maximized Performance at Minimized Vibratory Loads

Marilena D. Pavel[1]

[1] Faculty of Aerospace Engineering, Delft University of Technology, The Netherlands

1. INTRODUCTION

Rotorcraft (helicopters and tiltrotors) are generally reliable flying machines capable of fulfilling missions impossible with fixed-wing aircraft, most notably rescue operations. These missions, however, often lead to high and sometimes excessive pilot workload. Although high standards in terms of safety are imposed in helicopter design, studies show that *"it is ten times more likely to be involved in an accident in a helicopter than in a fixed-wing aircraft"* (Iseler et. al. 2001). According to World Aircraft Accident Summary (WAAS, 2002), nearly 45 percent of all accidents of single-piston helicopters is attributed to pilot loss of control, where - because of various causes, often involving vibrations, high workload, and bad weather - a pilot loses control of the helicopter and crashes, sometimes with fatal consequences. Quoting the Royal Netherlands Air Force ('Veilig Vliegen' magazine, 2003), *"for helicopters there is a considerable number of inexplicable incidents (...) which involved piloting loss of control"*. The situation is likely to get worse, as rotorcraft missions are becoming more difficult, demanding high agility and rapid manoeuvring, and producing more violent vibrations. (Kufeld & Bousman, 1995; Hansford & Vorwald, 1996; Datta & Chopra, 2002)

The primary cause of pilot control difficulties and high-workload situations is that even modern helicopters often have poor Handling Qualities (HQs) (Padfield, 1998). Cooper and Harper (Cooper & Harper, 1969), pioneers in this subject, defined these as: *"those qualities or characteristics of an aircraft that govern the ease and precision with which a pilot is able to perform a mission"*. Below, the current practice in rotorcraft handling qualities assessment will be discussed, introducing the key problem addressed in this chapter.

1.1. STATE-OF-THE-ART IN ROTORCRAFT HANDLING QUALITIES – THE AERONAUTICAL DESIGN STANDARD ADS-33

Helicopter handling qualities used to be assessed with requirements defined for fixed-wing aircraft, as stated in the FAR (civil) and MIL (military) standards. In the 1960's, however, it became clear that these standards were not sufficient (**Key, 1982**). Helicopters have *strong cross-coupling effects* between longitudinal and directional controls, their behaviour is *highly non-linear* and requires *more degrees of freedom* in modelling than the rigid-body models used for aircraft. Therefore, the MIL-H-8501A standard (**MIL-H-8501A, 1962**) was developed. This standard was used up until mid 1980's. From a safety perspective, these requirements were merely 'good minimums', and a new standard was developed in the 1970's, that is used up until today, the Aeronautical Design Standard ADS-33 (**ADS-33, 2000**)

The crucial point, understood by ADS-33, is that helicopter HQ requirements need to be related to the mission executed, as this will determine the needed pilot effort. E.g., a shipboard landing at night and in high sea with strong ship motions demands more precision of control from the pilot than when flying in daytime and good weather.2 ADS-33 introduced handling qualities *metrics* (HQM), a *combination* of flight parameters such as rate of climb, turn rate, etc., that reflect how much manoeuvre-capability the pilot has *per specific mission*. These metrics are then mapped into handling qualities criteria (HQC) that yield boundaries between 'good' (Level 1), 'satisfactory' (Level 2) and 'poor' (Level 3) HQs.[1] -

Despite their importance for the helicopter safety, its operators and, above all, the helicopter pilots, achieving good handling qualities is still mainly a secondary goal in helicopter design. The first phase in helicopter design is the 'conceptual design' phase in which the main rotor and fuselage parameters are established, based on desired performance and, to some extent, vibration criteria.20 Only in the following phase, that of preliminary design, are 'high-fidelity' simulation models developed and the handling qualities considered. The high fidelity models allow an analysis of helicopter behaviour for various flight conditions. Applying the ADS-33 metrics/criteria to these models results in predicted levels of HQs. When these are known, the *experimental* HQ assessment begins, illustrated in Fig. 1.

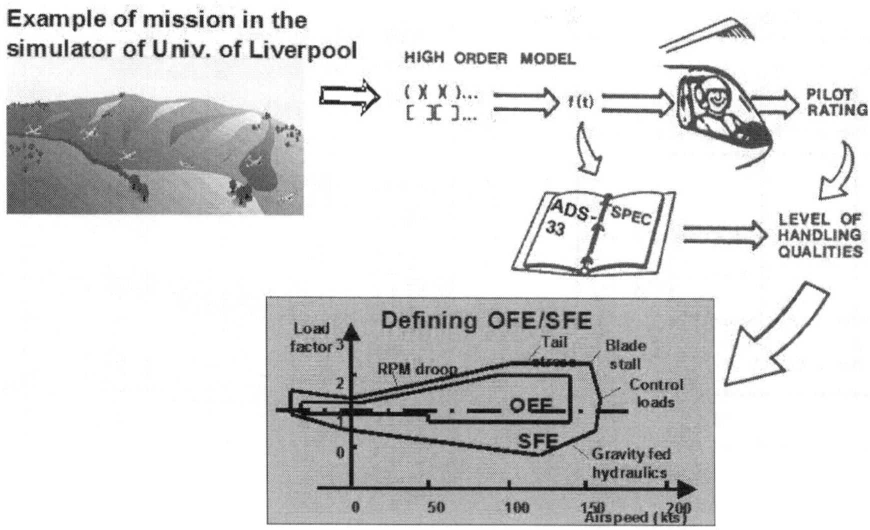

FIGURE 1. Experimental assessment of helicopter handling qualities

In this process, first databases of missions and environments are defined, according to customer requirements. Manoeuvrability, i.e., how easy can pilots guide the helicopter, and agility, i.e., how quickly can they change its flight direction, are critical performance demands. Second, experienced test pilots are asked to fly the missions in a flight simulator. Through insisting pilots to execute the mission very thoroughly, one aims to expose deficient handling qualities. Pilots rate the HQs they experience in each manoeuvre flown, generally using the Cooper-Harper rating scale.[2] - Third, Operational Flight Envelopes (OFEs) and Service Flight Envelopes (SFEs) are defined, based on a mapping of the HQ ratings. OFEs represent the limits within which the helicopter must be able to operate in order to accomplish the operational missions. SFEs stem from the helicopter limits and are expressed in terms of any parameters believed necessary to ensure safety, see Fig. 1.

In this first experimental assessment of the helicopter's handling qualities, the first problems arise, as more often than not, large differences arise between the theoretical predictions and the experimentally-determined pilot judgments. The gaps that occur are bridged by applying optimisation techniques using the simulation models developed in preliminary design, to improve the designs of helicopter, load alleviation system and flight control system (Celi, 1991; Celi, 1999; Celi, 2000; Fusato & Celi, 2001,Fusato & Celi, 2002a,2002b; Ganduli, 2004; Sahasrabudhe & Celi, 1997). Computational approaches have three important disadvantages, however. First, many designs that "roll out" of the procedure are unfeasible, the optimisation "pushing" the solution along the boundaries of the problem and not inside of the feasible region. Second, optimising for ADS-33 requires calculations of the helicopter time-domain responses, and the numerical methods become computationally very intensive (Sahasrabudhe & Celi, 1997; Tischler et. al., 1997) Third, and most important, the lack of quantitative, validated helicopter pilot models, capable of accurately predicting the effects of helicopter vibrations on pilot control behaviour, prevents the proper inclusion of pilot-centred considerations in mathematical optimization techniques (Mitchell et. al., 2004; Tischler et. al., 1996)

Whereas the first and second disadvantages are common in multi-dimensional design problems, the lack of knowledge on how helicopter vibrations affect pilot performance is a typical and fundamental problem of modern helicopter design (Mitchell et. al., 2004, Padfield, 1998). Although ADS-33 proposes criteria and missions regarding helicopter limits, these only characterize a helicopter's performance, and do not require an adequate knowledge of helicopter vibratory loads (Kolwey, 1996;Tischler et. al., 1996). This shortcoming stems from the fact that, when the ADS-33 criteria were defined, helicopter missions were not so demanding, and the vibratory loads associated with them were low. In the last twenty years, however, ever-increasing performance requirements and extended flight envelopes were defined, for reasons of heavy competition, demanding manoeuvres that impose heavy vibrations on both structure and pilot. These vibrations, combined with cross-coupling effects, rapidly lead to pilot overload and degradation in performance (Padfield, 2007).

Key Problem: The current practice of assessing rotorcraft handling qualities reveals significant gaps in the ADS-33 HQ criteria, especially regarding the effects of vibrations.

1.2. GOAL

The design of high-performance rotorcraft has become an arduous process, regularly leading to surprises, demanding 'patches' to safety-critical systems, and needing more iterations than expected, all contributing in very high costs. Unmistakably, the helicopter and flight control system design teams do not have up-to-date criteria to adequately assess the effects of helicopter vibrations on its handling qualities. There is an urgent need for a much more fundamental understanding of how helicopter vibrations affect pilot control behaviour (Mitchell et. al., 2004) and for new tools to incorporate this knowledge as early as possible in the design of both helicopter and flight control system. The goal of this chapter is to portray a novel approach to rotorcraft handling qualities (HQs) assessment by defining a set of consistent, complementary metrics for agility and structural loads pertaining to vertical manoeuvres in forward flight. These metrics can be used by the designer for making trade—offs between agility and vibrational/load suppression. The emphasis of the chapter will be on agility characteristics in the pitch axis applied to helicopter and tiltrotor. Especially in such new configurations, the proposed approach could be particularly useful as the performance tools for fixed-wing mode and helicopter mode must merge together within new criteria (Padfield, 2008).

The chapter is structured as follows: The second section will present an overview of traditional metrics for measuring pitch agility; The third section will present some alternative metrics proposed in the 90's for better capturing the transient characteristics of the agility; Then, based on the rational developments of the metrics from the previous two sections, fourth section will propose the new approach that can better quantify the agility from the designer point of view. Finally, general conclusions and potential extension of this work will be discussed.

2. TRADITIONALLY DESIGN OF AIRCRAFT FOR PITCH AGILITY

One of the most important flying quality concepts defining the upper limits of performance is the so-called "agility". Generally, it is well known that the level of performance achieved by the pilot depends on the task complexity. Fig. 2 presents generically this situation, showing that there is a line of saturation up to which the pilot is able to perform optimally the specified mission; increasing the task difficulty above this line leads quickly to stress, panic and even incapacity to cope anymore with the task complexity and blocking, sometimes with fatal consequences.

It is difficult to point precisely to the origins of the concept of agility but probably these go back to the moment when it was realized that, in a combat, a "medium performance" fighter could win over its superior opponent if the first aircraft possesses the potential for faster transient motions, i.e. superior agility. In its most general sense, the concept of agility is defined with respect to the overall combat effectiveness in the so-called "Operational agility". Operational agility according to measures the '*ability to adapt and respond rapidly and precisely, with safety and poise, to maximize mission effectiveness*'(McKay, 1994). In the mid 80's a strong wave of interest arose in seeking metrics and criteria that could quantify the aircraft agility (Mazza, 1990; McKay, 1994). However, there have been developed almost as many criteria of agility as there were investigators in the field. The problem was partially due to the lack of coordination in the research studies performed but also due to a disagreement on the most fundamental level: there simply was very little agreement on what agility was.

Within the framework of operational agility one can see agility as a function of the airframe, avionics, weapons and pilot. Airframe agility is probably the most crucial component in the operational agility as it is designed in from the onset and cannot be added later. The present chapter focuses on airframe agility and within this, the chapter will relate to the airframe agility in the pitch axis.

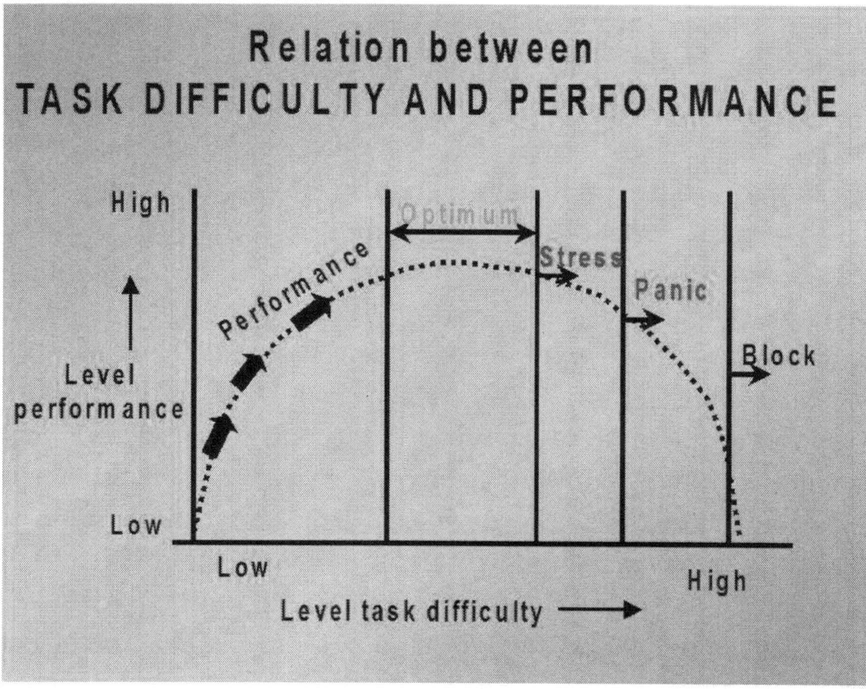

FIGURE 2. Correlation between task difficulty and performance

A large number of agility metrics have been proposed during the years to determine the aircraft realm of agility. The AGARD Working Group 19 on Operational Agility (McKay, 1994) put together all the different metrics and criteria existing on agility and fit them into a generalizable framework for further agility evaluations. The present section presents the traditional approach on pitch agility using as example a tiltrotor aircraft. This specific aircraft combines the properties of both fixed and rotary-wing aircraft and can be used to define a unified approach in the agility requirements at both fixed and rotary-wings.

The tiltrotor example to be investigated in this study is the Bel XV-15 aircraft. Next, as vehicle model, the FLIGHTLAB model of the Bell XV-15 aircraft as developed by the University of Liverpool (this model is designated as FXV-15) will be used. For a complete description of this model and the assumptions made the reader is referred to (Manimala et. al., 2003). For the tiltrotor in helicopter mode, the pilot's controls command pitch through longitudinal cyclic, roll through differential collective (lateral cyclic is also provided for trimming), yaw through differential longitudinal cyclic and heave through combined collective. In airplane mode, the pilot controls command conventional elevator, aileron and rudder (a small proportion of differential collective is also included).

Pitch agility is the ability to move, rapidly and precisely, the aircraft nose in the longitudinal plane and complete with easiness that movement. This implies that in the agility analysis one has to search for sample manoeuvres to be carried out by the flight vehicle dominated by high flight path changes and high rate of change of longitudinal acceleration which can give a good picture of the agility characteristics. As starting point in this discussion on pitch agility we will consider the kinematics of a sharp pitch manoeuvre – a simple example of this type is the tiltrotor trying to fly over an obstacle (seeFig. 3). Assume that the manoeuvre is executing starting from different forward speeds (helicopter mode 60 kts and 120kts; airplane mode 120 kts and 300 kts) and the manoeuvre aggressiveness is varied by varying the pulse duration (from 1 to 5 sec). The pilot flies the manoeuvre by giving a pulse input in the longitudinal cyclic stick of 1 inch amplitude.

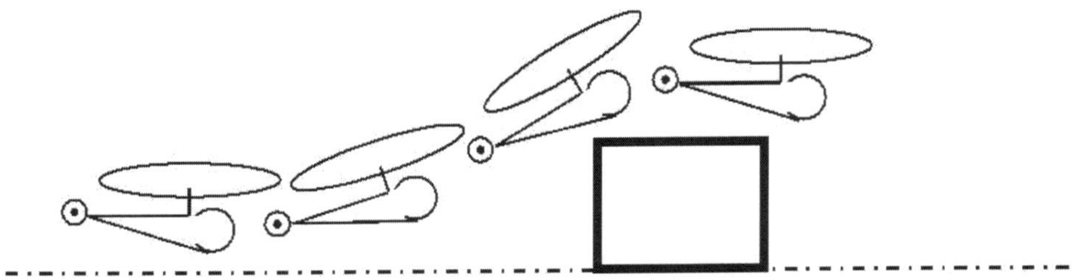

FIGURE 3. Executing an obstacle-avoid manoeuvre in the pitch axis

2.1. TRANSIENT METRICS

The first class of metrics developed to quantify the agility corresponds to the so-called "transient metrics". The transient class contains metrics which can be calculated at any moment for any manoeuvre. For pitch agility these metrics are pitch rate (entitled attitude manoeuvrability metric) and accelerations along the axes a_x, a_y, a_z (entitled manoeuvrability of the flight path). These metrics are next studied for the pull up manoeuvres flown with the FXV-15 in a 1 second pulse given from the initial trim at 120kts in helicopter mode and 300 kts in airplane mode. The presentation of the transient metric information is best achieved through a time history plot. Fig. 4 presents the transient metrics parameters of pitch rate q and vertical acceleration n_z (in the form of normal load factor). Looking atFig. 4 one may see local maxima in the metric parameters q and n_z illustrating peak events in the agility characteristics. This clearly demonstrates that in a "real" manoeuvre sequence, the agility characteristics occur at key moments, depending on the manoeuvre.

2.2. EXPERIMENTAL METRICS

The above conclusion gave the idea to develop a new class of agility metrics, the so-called "experimental metrics" formulated as discrete parameters during a real manoeuvre sequence. These metrics are actually the basic building blocks for understanding the agility and can be related to flying qualities and aircraft design. The metrics describing pitch agility during aggressive manoeuvring in vertical plane were defined by (Murphy et. al., 1991) and are described in the next section. They referred to the ability of an aircraft to pint the nose in at an opponent and commented that what was not clear in such manoeuvres was the behaviour of the flight path. Was the nose pointing w.r.t. the velocity vector or did it include the flight path bending or perhaps both? The authors noted that longitudinal stick displacements would be expected to command the flight path in addition to the aircraft nose pointing pitch angle for agile aircraft. The study pointed out that current aircraft behave differently in the high speeds and slow speed regimes. In the high speed case the flight path displaced as per the nose pointing displacement. The low speed case exhibited no flight path or even opposite flight path displacements.

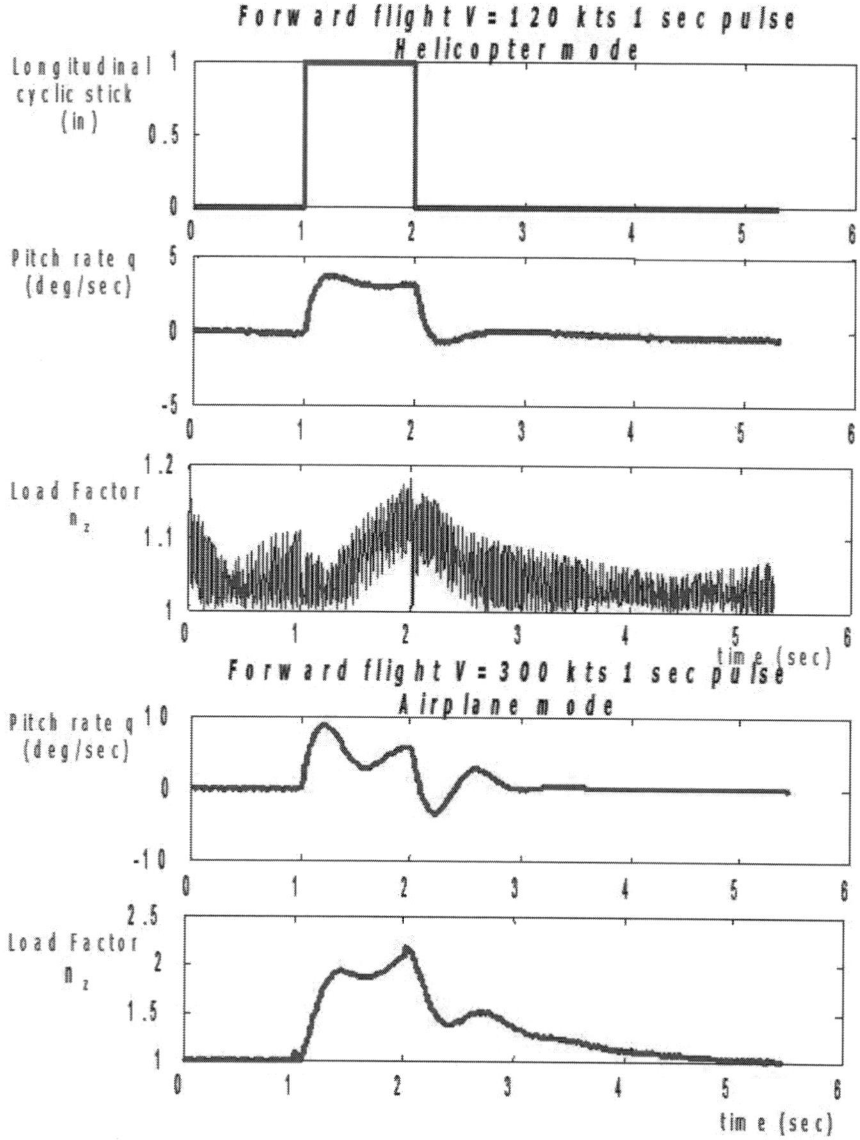

FIGURE 4. Transient agility metrics for pull-up manoeuvres with the tiltrotor

2.2.1. PEAK AND TIME TO PEAK PITCH RATES

The peak and time to peak pitch rates metrics were proposed by (Murphy et. al., 1991) for fixed wing aircraft. These metrics measure the time to reach peak pitch rate and the corresponding pitch rate. Fig. 5presents charts of peak pitch rate and time to reach this peak as a function of the velocity for the tiltrotor flying pull-up manoeuvres of increasing pulse duration. The pull-up manoeuvres are executed gradually increasing the velocity and the nacelle angle from the helicopter mode (90deg nacelle, hover and 60 kts) to conversion (60deg nacelle 120 kts) and ending in airplane mode (0deg nacelle 200 kts). Looking at these figures one can see that as the velocity increases the pilot is able to achieve higher pitch rates, the time to achieve these peaks being and faster especially if the pulse duration is short. As attributes, the peak and time to peak pitch rates metrics have the advantage that can be related to design.

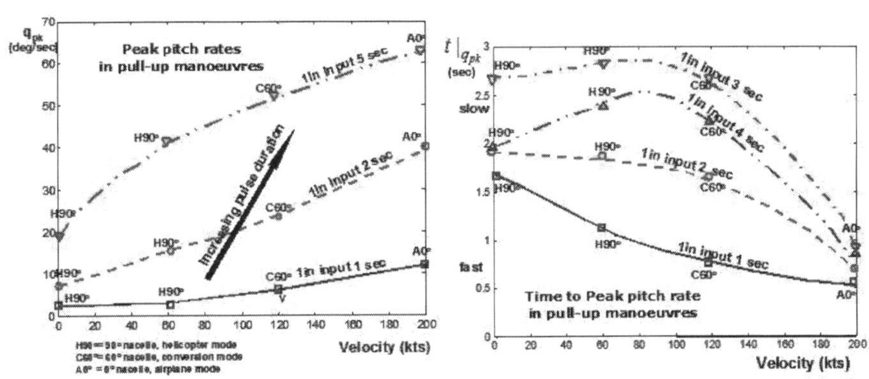

FIGURE 5. Peak and time to peak pitch rates in pull-up manoeuvres

2.2.2. PEAK AND TIME TO PEAK PITCH ACCELERATIONS

(Murphy et. al., 1991) considered the so-called peak and timer to peak pitch accelerations as the primary metrics for pitch motion agility. The time to peak acceleration provides insight into the jerk characteristics of pitch motion: if it is too slow, then the pilot may complain that the aircraft is too sluggish for tracking-type tasks; if it is too fast, then the pilot may complain of jerkiness or over-sensitivity. Fig. 6 presents charts of peak and time to leak pitch acceleration as a function of velocity when flying pull-ups manoeuvres. One can see that as the velocity increases the pilot is able to obtain higher pitch accelerations but as is passing from the helicopter to aircraft mode this capability diminishes. For fixed wing aircraft, (Murphy et. al., 1991) commented on the differences in the data for the peak accelerations in the body and wind axes. This effect has implications on the pilot selection of flight path or nose pointing control during manoeuvring.

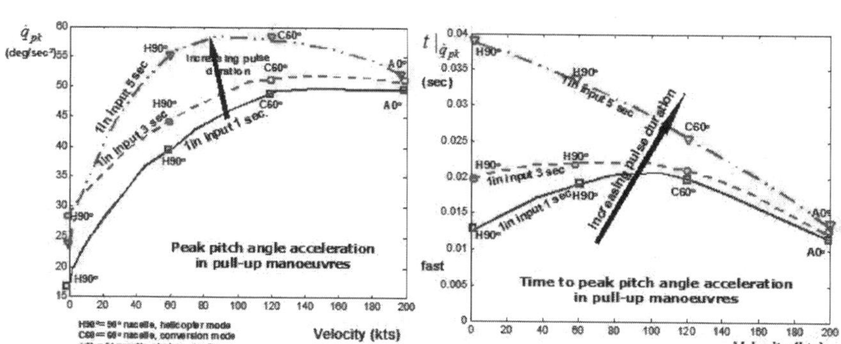

FIGURE 6. Peak and time to peak pitch angle acceleration in pull-ups with the tiltrotor

2.2.3. PEAK AND TIME TO PEAK LOAD FACTOR

Peak and time to peak load factor metrics describe the peak and the transition time to the peak normal load factor during a manoeuvre in pitch axis. They can be used at best to determine the flight path bending capability of an aircraft. Fig. 7 presents these two metrics as a function of the velocity for the tiltrotor example. One may see that as the velocity increases the pilot is able to pull more g's as going from the airplane to helicopter mode.

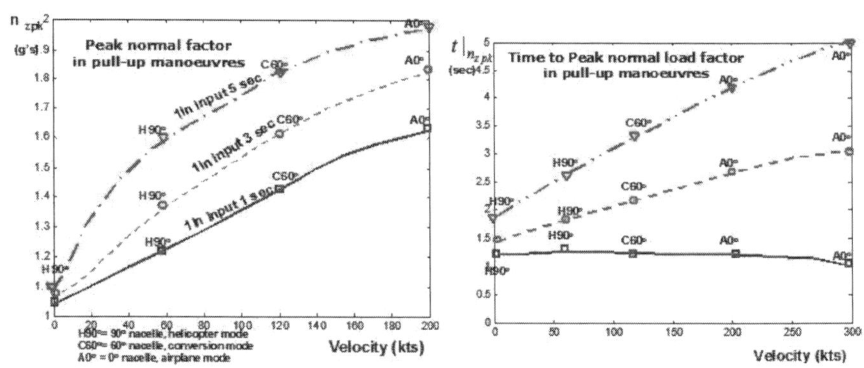

FIGURE 7. Peak and time to peak normal load factor

2.2.4 PITCH ATTITUDE QUICKNESS PARAMETER

One of the most important pitch agility metrics introduced by ADS-33 helicopter standard (**ADS-33, 2000**) is the so-called "pitch attitude quickness" parameter and is defined as the ratio of the peak pitch rate to the pitch angle change:

$$Q_\theta \stackrel{def}{=} \frac{q_{pk}}{\Delta\theta}\left(\sec^{-1}\right) \qquad (1)$$

The advantage of this parameter is that it was linked to handling qualities so that potential bounds for agility could be identified. In this sense, ADS-33 presents HQs boundaries for the pitch quickness parameter as a function of the minimum pitch angular change $\Delta\theta_{min}$ (considered as the pitch angle corresponding to a 10% decay from q_{pk}). These boundaries are defined to separate different quality levels, but because they relate too to an agility metric, they become now boundaries of available agility.Fig. 8 illustrates the attitude quickness charts for the tiltrotor executing pull-up maneuvers of 1 to 5 sec 1in amplitude input at 60, 120 and 300 kts in helicopter and airplane mode. The figure shows also the Level 1/2 boundaries as defined by 1) ADS-33 for a general mission task element, low speed helicopter flight (<45kts) and 2) MIL STD 1797A for fixed wing aircraft. One may see that whereas in helicopter mode FXV-15 hardly meets Level 1 performance in ADS-33 standard, being mostly at Level 2 performance, in airplane mode FXV-15 meets Level 1 performance in AHS-33 but exhibits Level 2 performance according to the MIL standard for airplanes (MIL HDBK-1797, 1997).

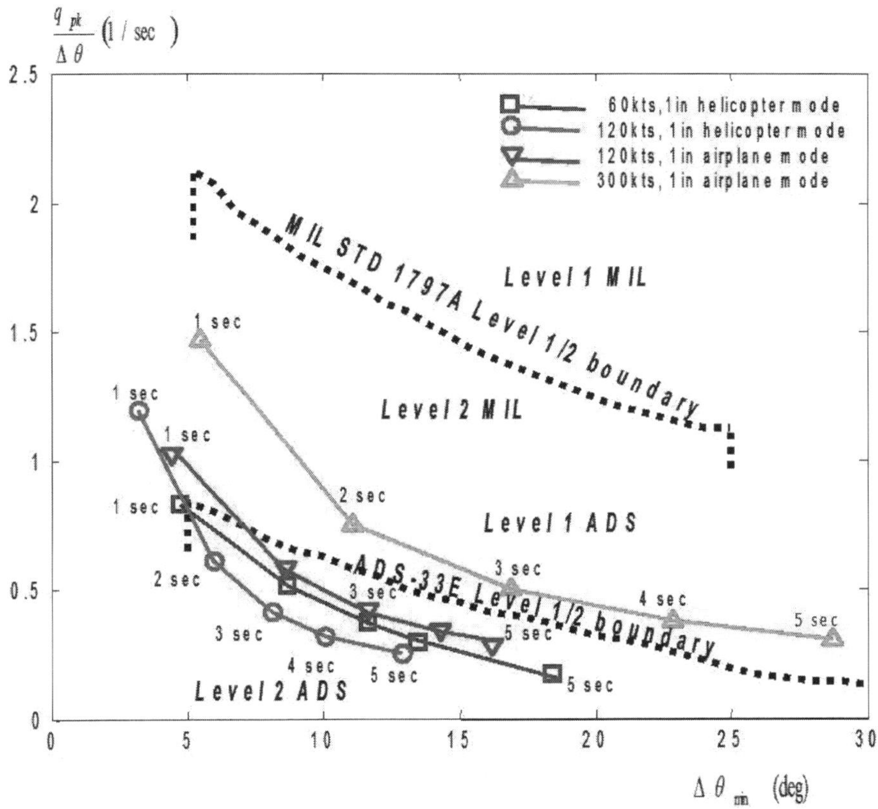

FIGURE 8. Pitch quickness for the tiltrotor

3. FLYING QUALITIES METRICS FOR AGILITY DESIGNING

Linking the agility to flying qualities raised up a new question: is agility limited by pilot handling parameters or in other words what are the upper limits to agility set by flying qualities considerations? Flying qualities considerations do limit agility according to (Padfield, 1998). In this sense, in a series of flight and simulation trials research conducted at DERA (now Qinetics) the pilots were asked to fly maneuvers with increasing tempo until either performance or safety limit was reached. The results showed that in all cases the safety limit came first, thus the agility was constraint by safety.

3.1. AGILITY FACTOR

A new metric was therefore introduced as a measure of performance margin (**Padfield & Hodkinson, 1993**), the so-called agility factor A_f, defined as the ratio of used to usable performance. For the simple case of the pull-up maneuver this metric can be easily calculated as the ratio of ideal task time T_i to actual task time T_a.

$$A_f \stackrel{def}{=} \frac{T_i}{T_a} = \frac{\omega_m \Delta t}{\omega_m \Delta t - \ln(0.1)} \quad (2)$$

where $T_i = \Delta t$ is the control pulse duration (1 to 5 sec), Ta is the time to reduce the pitch angle to 10% of the peak value achieved and ωm is the fundamental first-order break frequency or pitch damping which for this simple case represents the maximum achievable value of quickness. Fig. 9 illustrates the variation of Af with $\omega_m \Delta t$ -thus the quickness. The values considered for ω_m were: ωm=1.81 rad/s in hover helicopter mode, ω_m=2.6 rad/s at 60kts helicopter mode, ω_m=3.6 rad/s at 120kts 60deg conversion mode. Fig. 9 underlines an important aspect of the link between handling and agility: the higher the quickness, the higher the agility but when this agility is connected to Fig. 8 one may see that at the highest agility poor Level 2 ratings are awarded, i.e. the performance degrades rather than improves. This shows that actually, in practice, the closer the pilot flies to the performance boundary the more difficult it becomes to control the maneuver and thus the higher the agility the worse the HQs. In conclusion, handling qualities considerations do limit the agility.

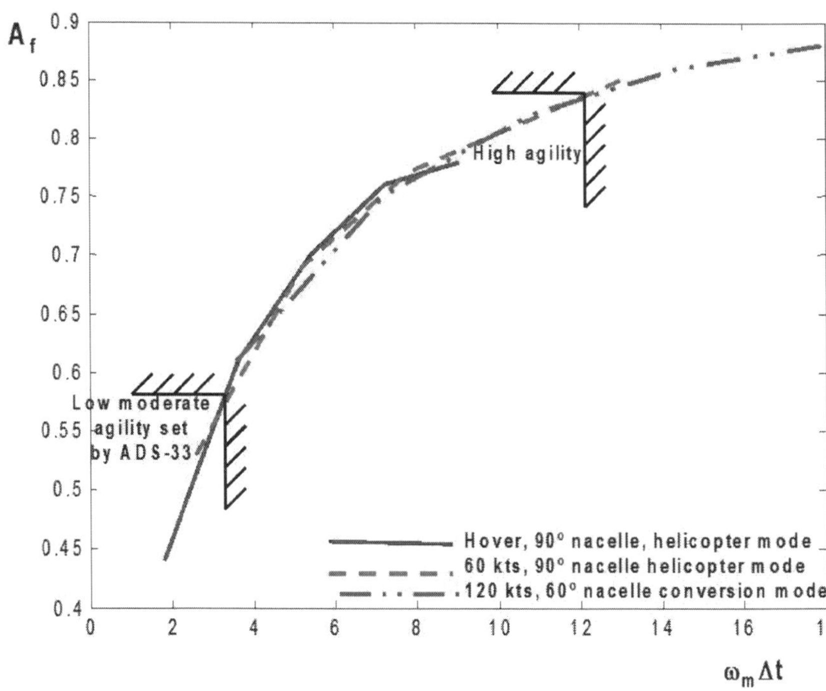

FIGURE 9. Agility factor as a function of quickness

3.2. CONTROL ANTICIPATION PARAMETER

The discussion on the experimental metrics suggests that on the one side the best metric for pitch motion agility is the peak pitch acceleration and on the other side the best metric for determining the aircraft flight path bending capability is the peak load factor. In order to capture both the transients of the maneuver and the precision achieved in flight path control, MIL standard on fixed-wing aircraft (MIL-HDBK-1797, 1997) introduced as metric a combination between these two metrics, the so-called 'control anticipation parameter CAP'. CAP is defined as the ratio of the initial pitch acceleration to the steady state load factor (effectively pitch rate) after a step-type control input:

$$CAP \overset{def}{=} \frac{\dot{q}(0)}{n_z^{qs}} \qquad (3)$$

MIL standard defines CAP boundaries for fixed-wing aircraft. Fig. 10 presents the agility of the FXV-15 CAP as a function of speed (60 kts, 120 kts and 200 kts) in the MIL boundaries. Looking at this figure one can see the tiltrotor meets Level 1 MIL performance and some degradation to Level 2 is seen when flying at high speeds in airplane mode.

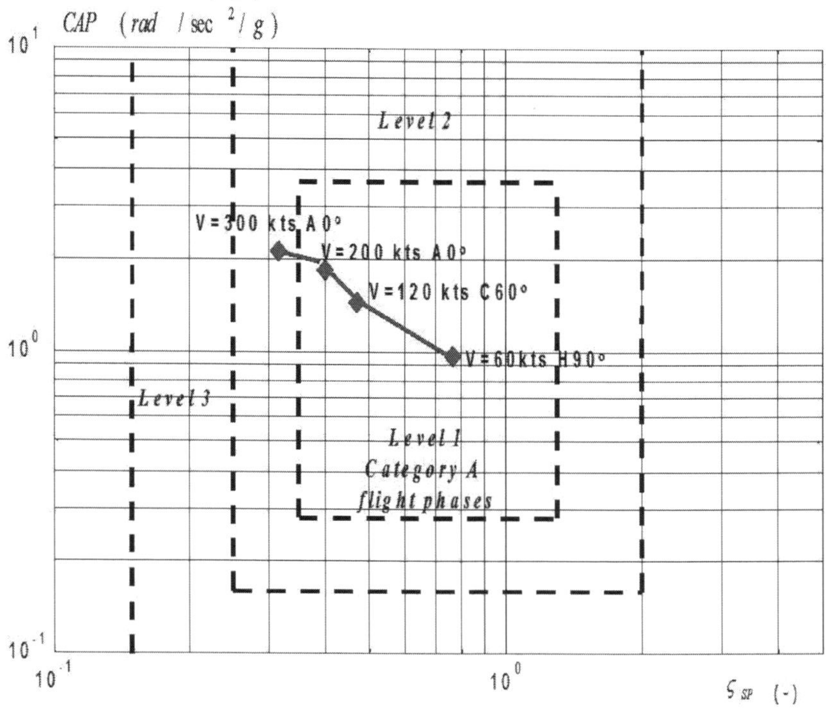

FIGURE 10. CAP boundaries for the tiltrotor

3.3. RATE PITCH QUICKNESS

For helicopters a similar metric to CAP was introduced by (Padfield & Hodkinson, 1993). This metric was called 'rate pitch quickness' and was defined as the ratio of pitch acceleration to the pitch angle change:

$$\dot{Q}_\theta \stackrel{def}{=} \frac{\dot{q}_{pk}}{\Delta\theta}\left(\sec^{-2}\right) \tag{4}$$

and can be used to determine upper limits to agility based on maneuver acceleration. Fig. 11 plotted the rate quickness in the normalized form as a function of acceleration time constant $\omega_m t_{pk}$ (where t_{pk} is the time to peak acceleration).

One can see that as the rate quickness increases the time to peak that rate is decreasing, so the agility is increasing. However, (Padfield & Hodkinson, 1993) commented that simply increasing the agility in terms of acceleration rates would lead to over-responsiveness and thus decreasing in operational capability since an over-responsive vehicle would not be controllable. In this sense, also CAP was quoted as an example of a criterion defining over-responsiveness. Unfortunately, there were no boundaries defined in this chart, although it was mentioned that intuitively there are likely to be upper and lower bounds for this metric. '*Hard and fast may be as unacceptable as soft and slow, both leading to low agility factors*' (Padfield & Hodkinson, 1993).

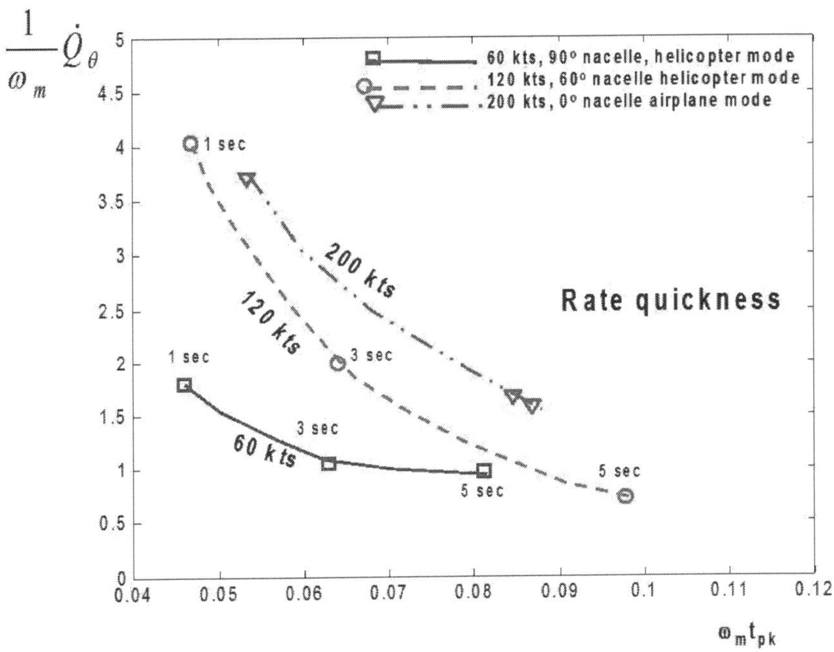

FIGURE 11. Rate quickness as a function of time peak acceleration

4. A RATIONAL DEVELOPMENT OF A MULTI-DISCIPLINARY APPROACH TO AGILITY

Combining equation (Eq. 3) for CAP with equation (Eq. 4) for rate quickness it follows that:

$$\dot{Q}_\theta = CAP \cdot \frac{n_z}{\Delta\theta} \tag{5}$$

Equation (5) gives the idea that rate quickness and CAP can be related to each other through a new metric which will be presented in the next paragraph.

4.1. AGILITY QUICKNESS METRIC AS A MEASUREMENT OF PERFORMANCE

As a potential successful metric for agility, (Pavel & Padfield, 2002) proposed a new metric for characterizing agility, the so-called 'agility quickness' defined as the ratio of peak quasi-steady normal acceleration $n_{z\,pk}^{qs}$ in g units corresponding to a step change in flight path angle Δγ:

$$Q_\gamma \overset{def}{=} \frac{n_{z\,pk}^{qs}}{\Delta\gamma} \quad \left(\frac{g's}{\deg}\right) \qquad (6)$$

Observe that the pitch angle from (Eq. 5) was substituted by the flight path angle, this has been done because actually during vertical axis maneuvering agility is more related to how quickly the flight path can be changed, the pilot being in reality more interested in the flight path angle change than in the pitch change. Furthermore, (Pavel & Padfield, 2003) proposed a Level 1/2 performance boundary for agility quickness by flying yo-yo maneuvers in the full motion simulator at the University of Liverpool the UH-60A model. Fig. 12 presents the example of tiltrotor on the agility quickness charts as determined in (Pavel & Padfield, 2003).

One can see that the tiltrotor is mostly at Level 2 performance in helicopter and airplane modes. (Pavel & Padfield, 2003) derived a relation between CAP and Q_γ and (Padfield & Meyer, 2003; Cameron & Padfield, 2010) connected CAP to other flying qualities parameters.

One of the reasons the attitude quickness criterion has gained large acceptance was due to its physical interpretation (in the limiting case gives the time constant of the aircraft as a function of the time constant of the maneuver). It can be demonstrated that agility quickness has also a physical interpretation, in the limiting case for small-amplitude maneuvers giving the heave damping, for large amplitudes giving the attitude quickness (Pavel & Padfield, 2002).

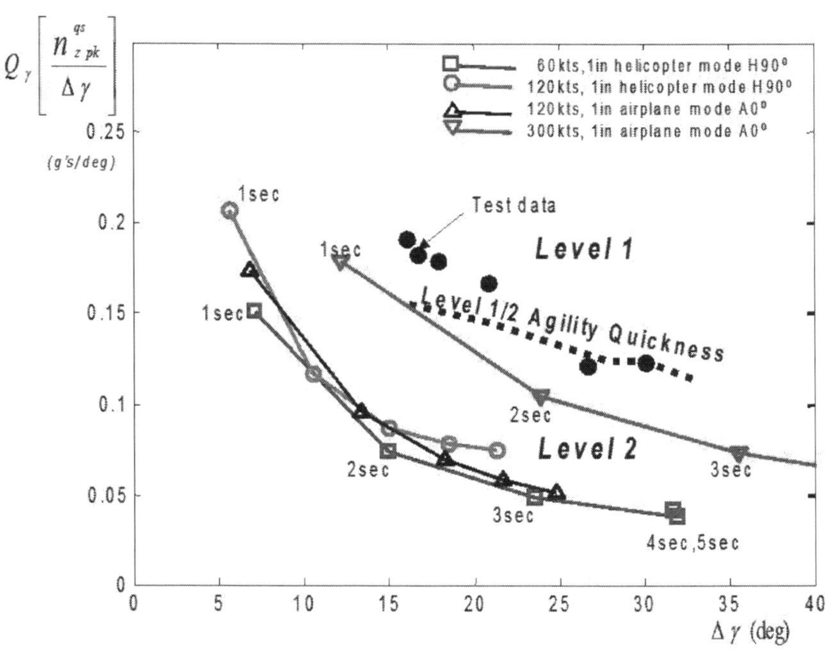

FIGURE 12. Tiltrotor on Agility quickness chart

4.2. VIBRATORY QUICKNESS METRIC AS A MEASUREMENT OF VIBRATORY ACTIVITY

The advantage of the agility quickness metric as above defined is that it could be linked to a complementary vibratory for structural alleviation. In this way, the designer is able to optimize in parallel both the performance and the vibratory loads in maneuvering flight. This novel approach may reform the methods presently used to accomplish agility of a new design as it is well known that, in practice, performance is itself often compromised by the need to control and minimize vibration.

(Pavel and Padfield, 2002) define an parallel metric, so-called 'vibratory load quickness' quantifying the buildup of loads in the rotor during maneuvering flight:

$$Q_l \overset{def}{=} \frac{F_{pk}^{vib}}{W\Delta\gamma}\left(\frac{1}{\deg}\right); \quad Q_l \overset{def}{=} \frac{M_{pk}^{vib}}{\Delta\gamma}\left(\frac{lbf \cdot ft}{\deg}\right) \tag{7}$$

Where F_{pk}^{vib}, M_{pk}^{vib} represent the peak amplitudes in the critical vibratory components for respectively hub shears and hub moments corresponding to a change $\Delta\gamma$ in flight path angle. The peak load amplitude can be calculated by using the FFT and time representations of the hub shears ($F_{x\ hub}$, $F_{y\ hub}$or $F_{z\ hub}$) and/or moments ($M_{x\ hub}$, $M_{y\ hub}$, $M_{z\ hub}$) during a maneuver flown and determining the critical loads (i.e. the loads achieving the highest peaks) during the maneuver.

For example, for the pull-up maneuver flown with the tiltrotor it was found that when flying in helicopter mode at 60 and 120 kts the critical loads developed were the 3/rev vibratory component of the hub vertical shear, the 1/rev and 2/rev components of the blade inplane moment and the 1/rev component of the blade flapping moment. When flying in the airplane mode at 120 and 300 kts, the critical loads measured by the FXV-15 were the 2/rev and 3/rev components of the vertical shear, the 1/rev and 2/rev components of the blade inplane moment.

Fig. 13 presents the equivalent vibratory quickness charts for the critical 3/rev component of the hub vertical shear in helicopter mode and 2/rev and 3/rev components of hub vertical shear in airplane mode when flying respectively at 60 and 120 kts and 120 and 300 kts, giving an 1 in input in longitudinal cyclic and varying the pulse duration (1 to 5 seconds). Each of these vibratory chart can be associated with an equivalent agility quickness chart as plotted in Fig. 12.

For helicopter mode it was plotted a presumable vibratory quickness boundary as derived in (Pavel & Padfield, 2003) when flying piloted yo-yo's in the simulator with the UH-60A helicopter. It was there showed that actually, increasing the flight path change enables the pilot to pull more g's of course but also increases the vibratory activity in the rotor. The presumable vibratory boundary would mean then that the structural designer would aim for a

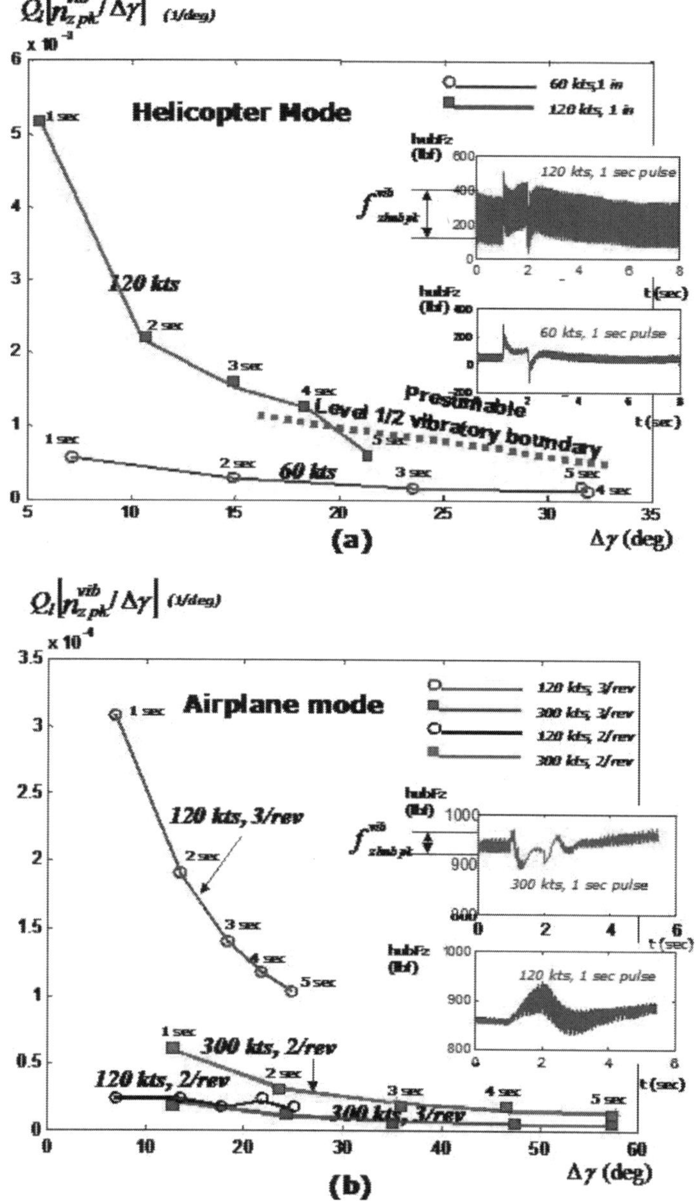

FIGURE 13. Vibratory quickness envelopes for the critical components of the hub vertical shear during a pull-up maneuver

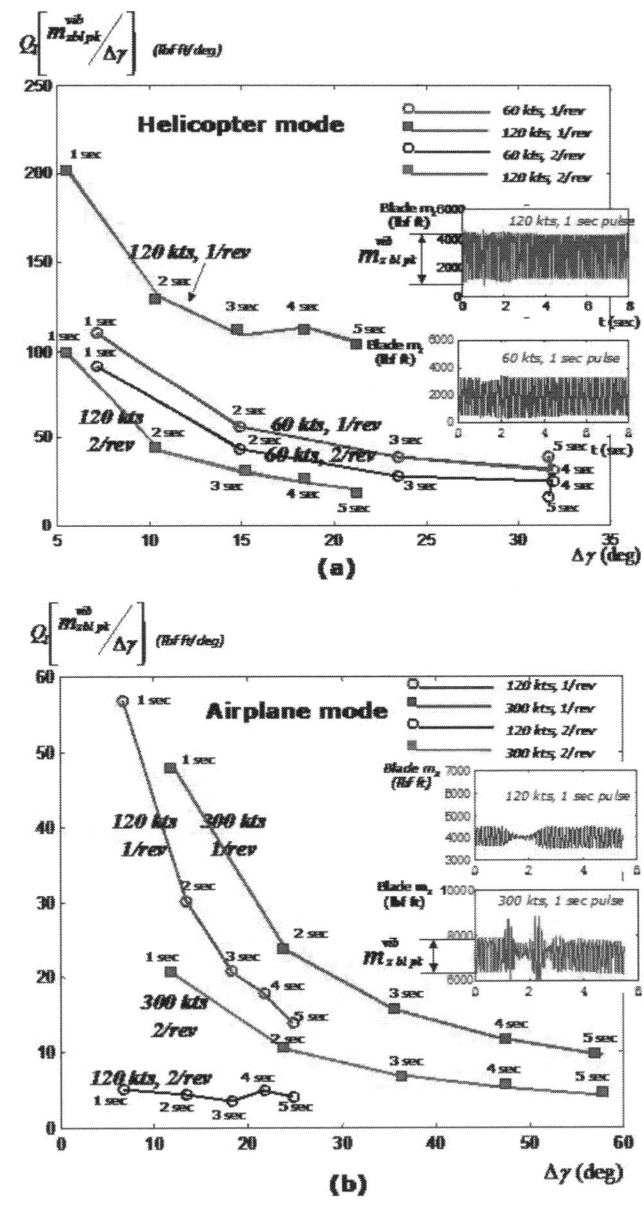

FIGURE 14. Vibratory quickness envelopes for the critical components of the hub vertical shear during a pull-up maneuver

boundary that slope in a similar direction to the agility quickness boundary as plotted in Fig. 13. Looking at this figure one may see that as the pulse duration is increasing the vibratory quickness is decreasing. This is because the vibratory activity in the hub reaches its absolute peak rather quickly, depending mainly on the initial velocity, the input amplitude (which is a measure of the level of aggressiveness in executing the maneuver) and not on the pulse duration. However, for very aggressive maneuvering, (Pavel & Padfield, 2002) showed that it might appear the situation in which, increasing the flight path change enabled the pilot to pull more g's (so, increased performance) but also increased the vibratory activity in the rotor. The goal of the structural designer would be then to alleviate these high peak loads to lower levels and reduce the sensitivity of the vibratory loads to flight path angle.

Fig. 14 presents other critical load for the tiltrotor, namely the blade inplane moment. Both, for the helicopter and airplane mode, the critical components measured during the simulation of the pull-up maneuvers with the FXV-15 were corresponding to the 1/rev and 2/rev vibratory components.

Looking at Fig. 14 one may see again that the vibratory quickness parameter Q_l varies approximatively inversely with the flight path change. This means that the vibratory activity in the blade in the inplane direction reaches its absolute peak rather quickly, depending on the aggressiveness of the pulse (pulse amplitude) and not on pulse duration.

5. CONCLUSION

This chapter has presented a rational development of key metrics and criteria used to design for airframe agility. Concentrating on the agility in the pitch axis (vertical-plane maneuvers) and taking as case study the unique example of a tiltrotor aircraft, the chapter demonstrated how, starting from the more traditional way of quantifying the agility, the designer can develop new agility metrics that do a better job of capturing the aircraft transient motion characteristics. This chapter discussed on the many correspondences that exist between the study of agility and the study of flying qualities, emphasizing the fact that flying qualities do limit agility. In this sense, providing the pilot with a high level of maneuverability, without a high level of controllability, will reduce agility. However, especially for the tiltrotor, higher agility cannot be achieved without increasing the vibratory loads on the rotor, which means also an increasing in pilot workload. The chapter proposed therefore a unique approach by presenting a first set of complementary metrics capable of being applied to both agility and structural load analysis.

Subsequent phases of this study will include the expansion of this new approach and studying it in axial, turning (horizontal-plane maneuvers) and roll (torsion) axes. It is hoped that in this way a more unified set of design criteria will be developed enhancing multi-disciplinary design optimization.

6. ACKNOWLEDGEMENTS

The subject of this research was initiated as collaboration between The University of Liverpool and Delft University of Technology. Sincere thanks for all support given by the University of Liverpool and Prof. Padfield for using FLIGHTLAB FXV-15 model.

NOTES

[1] - Level 1 HQs means that the rotorcraft is satisfactory without improvements required from the pilot's perspective. Level 2 HQs means that pilots can achieve adequate performance, but with compensation; at the extreme of Level 2, the mission is flyable, but pilots have little capacity for other duties and can not sustain flying for longer periods without the danger of pilot error. Level 3 HQs is unacceptable, it describes rotorcraft behaviour in extreme situations, like the loss of critical flight control systems.

[2] - The Cooper-Harper rating scale runs from 1 to 10. Rates from 1 to 3 1/2 correspond to Level 1 HQs; rates from 31/2 to 61/2 correspond to Level 2 HQs, and rates from 61/2 up to 10 correspond to Level 3 HQs.

REFERENCES

1. ADS-332000Aeronautical Design Standard-33E-PRF, Performance Specification, Handling Qualities Requirements for Military Rotorcraft, US Army AMCOM, Redstone, Alabama

2. Neil. Cameron, Gareth. D. Padfield, 2010Tilt Rotor Pitch/Flight-Path Handling QualitiesJ. Of the American Helicopter Society, 55(4), 042008042001to 042008-13

3. R. Celi, 1991Optimum Aeroelastic Design of Helicopter Rotors for Longitudinal Handling Qualities ImprovementJ. of Aircraft, 28(1), 4957

4. R. Celi, 1999Recent Applications of Design Optimization to Rotorcraft- A Survey, J. of Aircraft, 36 (1), pp.

5. R. Celi, 2000Optimization Based Inverse Simulation of a Helicopter Slalom Maneuver, J. of Guidance, Control and Dynamics, 23(2), 289297

6. G. E. Cooper, R. P. Harper, Jr , 1969The use of Pilot Ratings in the Evaluation of Aircraft Handling Qualities, NASA TM D-5133

7. A. Datta, I. Chopra, 2002Validation and Understanding of UH-60A Vibration Loads in Steady Flight, 58th American Helicopter Society Conference, 1114June, Montreal

8. D. Fusato, R. Celi, 2001Formulation of a Design-Oriented Helicopter Flight Dynamic Simulation Programth American Helicopter Society Forum, Alexandria, VA

9. D. Fusato, R. Celi, 2002Design Sensitivity Analysis for ADS-33 Quickness Criteria and Maneuver Loadsth American Helicopter Society Forum, 1114June, Montreal

10. D. Fusato, R. Celi, 2002Multidisciplinary Design Optimization for Aeromechanics and Handling Qualitiesth European Rotorcraft Forum, 1720September, Bristol, UK

11. R. Ganduli, 2004Survey of Recent Developments in Rotorcraft Design Optimization", J. of Aircraft, 41(3), 493510

12. R. E. Hansford, J. Vorwald, 1996Dynamic Workshop on Rotor Vibratory Loads. 52nd Forum of the American Helicopter Society, June 46Washington D.C.

13. Iseler, Laura; De Maio, Joe & Rutkowski, Michael2002Analysis of US Civil Rotorcraft Accidents by Cost & Injury (1990-1996), Army/NASA Rotorcraft Division Ames Research Center, USA, NASA-TM-2002209615

14. D. L. Key, 1982The status of Military Handling Qualities Criteria, AGARD CP-333

15. H. G. Kolwey, 1996The New ADS-33 Process: Cautions for ImplementationTechnical Note, J. of the American Helicopter Society, 41(1), 36

16. R. M. Kufeld, W. G. Bousman, 1995High Load Conditions measured on a UH-60A in Manoeuvring Flight, 51st Forum of the American Helicopter Society, May 911Texas

17. K. Mc Kay, 1994Operational Agility: An Overview of AGARD Flight Mechanics Working Group 19, In: Technologies for Highly Manoervrable Aircraft, AGARD-CP-548

18. Binoy. Manimala, Mauro. . Naddei, Philippe. Rollet, 2003Load Alleviation in Tiltrotor Aircraft through Active Control; Modelling and Control Concepts, 59th American Helicopter Society Conference, Phoenix, Arizona, May 6-8, 2003

19. C. J. Mazza, 1990Agility: A Rational Development of Fundamental metrics and their relationship to Flying Qualities. Proceedings of Flight Mechanics Panel Symposium, Quebec City, Canada, October 15-18, 1990, Published as AGARD-CP-508

20. MIL-H-8501A1962Military Specification, General Requirements for Helicopter Flying and Ground Handling Qualities MIL-H-8501A, Sept. 1961, through Amendment 1, April 1962 (superseding MIL-H-8501, Nov. 1952).

21. MIL-HDBK-17971997Flying Qualities of Piloted Aircraft, MIL-HDBK-1797, US Department of Defense, USA

22. D. G. Mitchell, et al.2004Evolution, Revolution and Challenges of Handling Qualities, J. of Guidance, Control and Dynamics, 27(1), 1228

23. P. Murphy, M. Bailey, Ostroff, 1991A. Candidate Control Design Metrics for an Agile Fighter, NASA-TM-4238

24. Gareth. D. Padfield, 1998The Making of Helicopter Flying Qualities; A Requirements Perspective. The Aeronautical Journal of Royal Aeronautical Society, 1021018409437

25. G. D. Padfield, 2007Helicopter Flight Dynamics, Blackwell Science LTD

26. G. D. Padfield, 2008Capturing requirements for tiltrotor handling qualities- case studies in virtual engineering, The Aeronautical Journal, 112Aug. 2008, 433448

27. Gareth. D. Padfield, John. Hodkinson, The Influence of Flying Qualities on Operational Agility", AGARD Flight Mechanics Panel Symposium Technologies for Highly Manoeuvrable Aircraft, Annapolis, Maryland, Oct. 1993Published as AGARD-CP-548

28. Marilena. D. Pavel, Gareth. D. Padfield, 2002Defining Consistent ADS-33Metrics for Agility Enhancement and

Structural Loads Alleviation, 58th American Helicopter Society Conference, Montreal, Canada, 11-14 June 2002

29. Marilena. D. Pavel, Gareth. D. Padfield, 2003Progress in the Development of Complementary Handling and Loading Metrics for ADS-33 Maneuvers", 59th American Helicopter Society Conference, Phoenix, Arizona, 68May 2003

30. Gareth. D. Padfield, Michael. Meyer, 2003Progress in Civil Tilt-Rotor Handling Qualities", 29th European Rotorcraft Forum, Friedrichshafen, Germany, 1517October 2003.

31. V. Sahasrabudhe, R. Celi, 1997Efficient Treatment of Moderate Amplitude Constraints for Helicopter Handling Qualities Design Optimization, J. of Aircraft, 34(6), 730739

32. M. B. Tischler, et.al, 1997CONDUIT- A New Multidisciplinary Integration Environment for Flight Control Development, NASA TM-112203

33. M. B. Tischler, J. N. Williams, Ham, J. A. , 1996Comments on The New ADS-33 Process: Cautions for Implementation, Technical Note, J. of the American Helicopter Society, 41194195

34. Veilig Vliegen magazine2003Hoe de zaken liepen in 2002' ('How the things evolved in 2002'), May 2003, 133138

35. WAAS2002World Aircraft Accident Summary-19902001Index-Cap 479, Civil Aviation Authority, Airclaims Limited

3

CHAPTER

PTVC-M for Ultra-Agile VTOL and 300+ km · h−1 Cruising

Chung-Kiak Poh*, **Chung-How Poh**

Aero-Persistence Research, Penang, Malaysia

ABSTRACT

There remains a need to develop improved VTOL techniques that are cost-effective and with mini- mum compromise on cruising flight performance for fixed-wing aircraft. This work proposes an elegant VTOL control method known as PTVC-M (pitch-axis thrust vector control with moment arms) for tailsitters. The hallmark of the approach is the complete elimination of control surfaces such as elevators and rudder. Computer simulations with a 1580 mm wing span airplane reveal that the proposed technique results in authoritative control and unique maneuverability such as inverted vertical hover and stall-spin with positive climb rate. Zero-surface requirement of the PTVC-M virtually eliminates performance tradeoffs between VTOL and high-speed flight. In this proof-of-concept study, the VTOL-capable aircraft achieves a

V_H of 360 km · h^{-1} at near sea-level. The proposed technique will benefit a broad range of applications including high-performance spinsonde that can directly measure 10-m surface wind, tropical cyclone research, and possibly serving as the cornerstone for the next-generation sport aerobatics.

Keywords: Pitch-Axis Thrust Vector Control, VTOL, Tailsitter, Ultra-Maneuverability

1. INTRODUCTION

In the early 1920s, Juan de la Cierva came up with the concept of autorotation in the efforts to solve the wing- stall problems [1] . His work and dedication to the development of autogyro have brought the notions of rotary wings and STOL (short takeoff and landing) to the aviation world [2] - [4] . Helicopters eventually emerged and since 1950s there have been quests for an aircraft that has the cruising efficiency of an airplane and the vertical takeoff and landing (VTOL) capability of a helicopter [5] . The Fairey Rotodyne has tip jet-powered rotor and two turboprops mounted under its fixed-wings [6] . The jet-powered rotor was driven for vertical takeoffs, landings and hovering, but autorotation mode was used during cruising with the forward thrusts coming from the turboprops [6] . It held a world speed record of 307.2 km · h^{-1} but the program was unfortunately terminated in 1962 [6] . Throughout the years numerous VTOL-capable airplanes have emerged and among them were the Convair XFY Pogo, the Bell XV-3, the Yakovlev Yak-141, and the Bell XV-15 [5] [7] . Despite significant efforts on the development of VTOL airplanes, numerous challenges remain. The primary challenges are often associated with the mechanical complexities, cost-of-ownership and maintenance [5] [8] . The ability to overcome these challenges will promote the widespread use of VTOL airplanes especially for civilian applications. Autonomous applications such as weather surveillance in remote oceans for an extended period of time would require a reliable airframe to help ensure high rate of mission success [9] . VTOL-capable airplanes can be broadly divided into several classes, namely tiltrotor, tiltwing, and tailsitter [5] . More recent developments include stop-rotors and hybrid [10] - [14] . In our opinion, apart from cost and reliability considerations, an important attribute that will significantly enhance the usefulness of a VTOL airplane is ultra-maneuverability, especially in the domain of unmanned aircraft.

Modern aerobatic model airplanes have the abilities to perform an array of impressive post-stall maneuvers known as "3D aerobatics" such as hovering (including VTOL), waterfall, flatspin, blender, tailslide and their derivatives [15] - [17] . The flight performance, and design reliability of these airplanes have been put to test globally during weekend flying in aeromodelling clubs and in international competitions such as the prestigious Fédération Aéronautique Internationale (FAI) Class F3M [17] . The distinctive characteristics of these aircraft are their relatively large control surfaces immersed in strong propeller wash and thrust-to-weight ratio that exceeds unity [15] . These techniques can potentially be used to create VTOL airplanes with ultra-agility. However, two major setbacks of this approach are that the control surfaces have to be relatively large, and that they should be immersed in propeller wash, and these limit the application scope. For example, if the technique were to be applied to an aircraft with anhedral (reversed dihedral) horizontal stabilizers with wing-mounted propellers, the pitch control authority has been found to be asymmetrical depending on whether the aircraft is performing the upright "harrier" (i.e. to fly in trim with nose angle of 45° or greater) or the inverted "harrier". Similarly, considerations have to be given to the placement of the rudder. The challenge is less pronounced when it comes to roll control as the aileron (s) can be suitably located almost anywhere along the trailing edge of each wing. Additionally, large control surfaces often pose a performance tradeoff during high speed cruising due to flutter [18] . Thrust vector control (TVC) methods have been used on aircraft, rockets, and multirotor UAVs (unmanned aerial vehicles) to control altitude or angular velocity [19] [20] .

To address the issues above in regard to the constraints imposed by the placement of rudder and elevators, this work proposes the concept of pitch-axis thrust vector control with moment arms (PTVC-M) in an attempt to create an ultra-agile 3-axis VTOL control method for tailsitters. The proposed PTVC-M requires no control surfaces to actuate VTOL flight controls so as to gain independence from the aerodynamic rudder and elevators. For an aircraft with twin-engine configuration (i.e., one engine on each wing), the system will involve two pairs of moment arms; one of which is parallel with the longitudinal axis to enable pitch control and the other pair is parallel with the lateral axis to ensure responsive yaw. The flight performance is investigated via computer simulation using a 1580 mm wing span UAV as a test platform. In this article, the definition of a tailsitter is one whereby the roll axis (longitudinal axis) of the aircraft is approximately parallel with the z-axis of the world frame during the VTOL phase though the landing gears can be attached to any part of the airframe including fuselage and wing and are not restricted to the vertical or horizontal stabilizers at the rear of the aircraft.

2. MATERIALS AND METHODS

2.1. Working Principle

In principle, the PTVC-M actuates pitch control by creating moments about the lateral axis (pitch axis) similar to conventional elevator effect, except the force is generated directly by the thrust-producing element instead of surface-derived aerodynamic force. In the case of propeller being the thrust-producing element, the moment arms extend from the center of the propeller disks to the lateral axis. In Figure 1, the pitching moment arm, L_1 is indicated for one set of wing. The yaw control of the PTVC-M is derived from the differential thrust of the propellers and a pair of moment arms. The yawing moment arm, L_2 is the distance from the center of the propeller disk to the centerline of the aircraft (Figure 1).

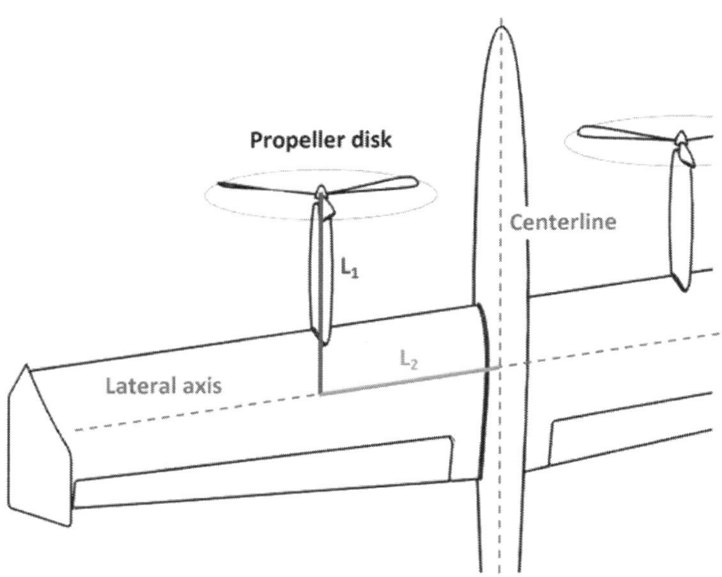

Figure 1. PTVC-M: illustration of the moment arms on one set of wing.

Therefore, the further away the location of the propeller from the centerline, the greater the maximum yaw moment. However, if the design intent requires the elevators to be partially immersed in the propeller wash to harness its additional advantages, then there should be an optimal placement of the propeller along the lateral axis because if placed too far out then there would be very little amount of propeller wash reaching the elevators at the rear of the fuselage. Apart from propeller, the thrust-producing element can include turbine fans, rotor blades, or possibly even rockets. The thrust-vectoring mechanism responsible for the tilting of the propeller disk can be realized using several well established methods such as swashplate or servo-actuated motor mounts similar to those being used for yaw control in present day tricopters [21] [22] . In the presence of downstream surfaces such as wing, the propeller wash effects [23] have to be considered. A way to avoid the propeller wash effect is to mount the thrust-producing elements towards the rear of the aircraft. However, such configuration may entail a more complex structural design. In this work, we will focus on a configuration in which the thrust-pro- ducing elements are mounted on the wing for the following reasons: 1) in the real-world scenario the wing spar would readily provide strong mounting points for the engine nacelle hence simplifying the structural design, and 2) control surfaces immersed in propeller wash will provide additional control authority and might allow a seamless transition in controls between VTOL and level flight.

2.2. Aircraft Design and Simulation Details

The graphical model of the small UAV used in this work was created using the Autodesk 3ds Max®[24] . The orientations of the pivots and naming convention of the components were as per the requirements and it was imported into the RealFlight®[1] 6.5 simulator [25] using the KEmax plugin [26] . Refinement of the computer model was done using the Accu-Model™ aircraft editor within the simulation software. The model UAV has a wing span of 1580 mm and a total body length of 1120 mm. The MH 22 airfoil [27] was used on the main wing to help achieve a high V_H (i.e. maximum speed in level flight with maximum continuous power). The model has twin motor, one on each side of the wing as shown in Figure 2. The airframe adopted the 3-piece stabilizers design, which consisted of a vertical stabilizer and two anhedral (reversed dihedral) horizontal stabilizers, spaced 120° apart from each other and they also functioned as part of the landing gears. Such configuration was expected to be lighter and has lower aerodynamic drags than the 4-piece stabilizers design commonly employed by tailsitters. The innovative "zero-surface" requirement of the proposed VTOL control technique enabled the designs of the aerodynamic rudder and elevator to be optimized for other phases of flight more critical to the primary roles of the aircraft such as high-speed cruising.

The surface areas of the rudder and elevators were approximately 32% of those of the stabilizers, which were significantly smaller than those on the "3D aerobatic" model aircraft [15] [16] . Maximum deflections of all the control surfaces were ±45°. Optional retractable undercarriage can be affixed to the UAV to enable takeoff and landing on runway. The aircraft was equipped with two 3-bladed propellers and they were counter-rotating to each other. The variable pitch propeller has a diameter of 305 mm. Each brushless motor has a maximum electrical power of 4 kW. The all-up weight (AUW) of the aircraft was 2.48 kg

Figure 2. Computer model of the twin-motor tailsitter incorporating the PTVC-M feature. The wing span is 1580 mm.

with a wing loading of 60.3 gdm^{-2}. Gyros (electronic flight stabilization system) were added to the roll, pitch and yaw controls for the piloted simulations in this study. However, flight stabilization for a particular principal axis would be disengaged when performing the angular rate test so as to avoid any possible influence from the gyro. Figure 3(a) to Figure 3(d) show the angular displacements of the propellers with the control inputs. To better illustrate the tilting of the propellers in relation to the control inputs, the displacements of the propellers were coupled to the control surfaces. The following describes the flight control methodology of the aircraft:

Pitch: tilting both the propellers in the same direction [Figure 3(a)].

Roll: tilting the propellers by the same angular displacement but in opposite direction [Figure 3(b)].

Yaw: differential thrust by varying the rotational speed of the propellers [Figure 3(c)]. In this case, right rudder was applied and the left-hand side propeller was spinning while the right-hand side propeller was stationary. The throttle was at its lowest position.

Figure 3(d) depicts the tilting orientations of the propellers (neglecting the propeller wash effects on the roll control) when up-elevator and right-aileron were applied simultaneously.

3. SIMULATION RESULTS AND DISCUSSION

3.1. Steady Level Flight Performance

The PTVC-M was deactivated and only the control surfaces were used during level flight evaluation, and the roll, pitch and yaw controls were found to be responsive. As shown in Figure 4, the maximum speed in level flight with maximum continuous power (V_H) was 360 ± 1 km · h^{-1} with total electrical power consumption of 6961 ± 1 W and the variometer indicated a vertical climb rate of 0 ms^{-1}. The power value displayed on the NavGuides in the upper left corner is the power consumption per motor. The stall speed (V_S) of the aircraft was determined to be 39 ± 1 km · h^{-1}. Any maneuvers performed at airspeeds below the V_S are termed as post-stall maneuvers in this work. When power to the left motor was cut off at cruising speed of 250 km · h^{-1}, the UAV yawed left and rolled left at rates of $14^\circ s^{-1}$ and $2^\circ s^{-1}$, respectively which were well within the capacity of the rudder and ailerons to counteract. As mentioned above, the gyros for these controls were switched off during the test.

Figure 3. Tilting orientations of the propellers with respect to the control inputs. Differential thrust was used to actuate yaw.

Figure 4. The tailsitter achieved a V_H of 360 km · h^{-1} with a total electrical power consumption of 6961 W.

3.2. VTOL and Agility Characterizations

Figure 5 shows the tailsitter in stationary vertical hover using only the PTVC-M without the control surfaces. Electrical power consumption was 222 W per motor. The hovering was stable and all the rotational controls around the principal axes were found to be authoritative. As the PTVC-M did not require control surfaces to work, it enabled the aircraft to perform inverted vertical hover, as shown in Figure 6. This would represent a true "four-dimensional" (4D) aerobatic flyer. The aircraft could comfortably perform an inverted vertical climb at a rate of 5 ms^{-1} before the aerodynamic moments from the surfaces overwhelmed those generated by the PTVC-M. The major aerodynamic difference between the upright and inverted hovers is that the former involved propeller wash effects due to the presence of downstream surfaces (e.g. wings and stabilizers). Subsequent characterizations in this section referred to the upright case, since it is the more likely configuration to be used for VTOL.

The maximum rates of the control response at near zero airspeed were 180°s^{-1}, 220°s^{-1} and >700°s^{-1} for the roll, pitch and yaw, respectively, but the effectiveness of PTVC-M dropped off with increasing airspeed. Strengthening of the PTVC-M yaw moment as the airspeed approached zero is similar to the fact that a multi-engine airplane with an engine failure will experience a larger yaw moment at low airspeed [28] .

A more comprehensive study on the control effectiveness of PTVC-M as a function of airspeed was carried out and the test method involved launching the tailsitter vertically on its own propeller thrusts. When the airspeed has reached the desired constant value, maximum input was applied to the roll. The test was evaluated at airspeeds of 60 and 150 km · h^{-1} and the process was repeated for the pitch and yaw (Table 1). While the

Figure 5. The tailsitter in a stationary hover using only the PTVC-M.

Figure 6. Inverted vertical hover made possible with the surface-free PTVC-M. Variometer shows a climb rate of 0.1 ms^{-1}.

roll rates can be regarded as equilibrium values, the rotational rates for the pitch and yaw were considered as quasi- state values before significant change in directional heading occurred which would alter the airspeed, and hence the rotational rates. The responsiveness of the control surfaces was evaluated using the same vertical launch procedure and they too were summarized in Table 1. Results indicated that the PTVC-M and the control surfaces are complementary, in that the control surfaces generally worked well at high speed, while the PTVC-M gave good response below 60 km · h^{-1}. This means that post-stall flights including VTOL could be primarily handled by the PTVC-M while the control surfaces could be optimized for other phases of flight thereby, eliminating the performance tradeoffs between VTOL and high speed cruise commonly encountered in VTOL airplane design.

The ailerons continued to work below the wing-stall speed of 39 km · h^{-1} because they were immersed in the strong propeller wash. The elevators managed to exert an angular rate of 50°s^{-1}at zero airspeed due to the widening of the wake flow angle during the static hover (0 km · h^{-1}airspeed) [23] . As expected, the aerodynamic rudder control completely stalled at zero airspeed due to lack of propeller wash in its vicinity.

The rolls of the simulated tailsitter were quite axial and have low tendency to cause directional change or promote angular momentum exchange among the principal axes of the aircraft [23] . Thus, it was possible to acquire the equilibrium roll rates attributed to the PTVC-M and the control surfaces for airspeeds ranging from 0 to 190 km · h^{-1}, as shown in Figure 7. The aileron roll rate was 230°s^{-1} at 0 km · h^{-1} and increased to 1215°s^{-1} at 190 km · h^{-1}. The PTVC-M, on the other hand, gave a slow roll rate of only 14°s^{-1} at 190 km · h^{-1}. Furthermore, Figure 7 suggests that the VTOL roll control could be adequately handled using only ailerons since it has good roll rates in the post-stall region (0 to 39 km · h^{-1}).

Setting the throttle to maximum while the aircraft was in hovering position and applying full deflection of pitch control resulted in a pitch rate of no less than 650°s^{-1}, which could be used to create a new form of maneuver based on the well-known "waterfall" maneuver [29] . This unique maneuver is available as a demonstration video (Video 1). The rapid yaw rate of the PTVC-M would benefit the development of high-performance spinsonde which will be covered in the next section of this article.

Video 2 shows the general VTOL and cruise flight performance of the tailsitter such as hovering, transition to forward flight (and vice versa), loop, and rolls. The maximum controlled vertical descent rate was found to be 8 ms^{-1}. The landing sequence from level flight to final touchdown could be performed with ease, an indication that the coupling between the control surfaces and the PTVC-M has resulted in responsive controls that seamlessly spanned across the full flight envelope.

3.3. Spinsonde Performance

Spinsonde is a concept developed to acquire high-resolution vertical wind speed profile measurements for atmospheric research [30] . It is a chute-free vertical retardation technique which enables multi-cycle measurements to be performed within a single flight and it is based on the stall-spin maneuver [30] . Multi-cycle measurement is not attainable with conventional expendable dropsondes [31] . This section explores the spinsonde capability of the aircraft whose primary flight controls involved the PTVC-M working in tandem with the aerodynamic surfaces.

Table 1. Comparison of angular rates about the principal axes due to PTVC-M and control surfaces at various airspeeds.

Control mode	Airspeed (km h^{-1})	Rotations about the principal axes (°s^{-1})		
		Roll	Pitch	Yaw
PTVC-M	0	180	220	>700
	60	70	110	35
	150	24	80	25
Control surfaces	0	230	50	0
	60	497	180	115
	150	1030	330	140

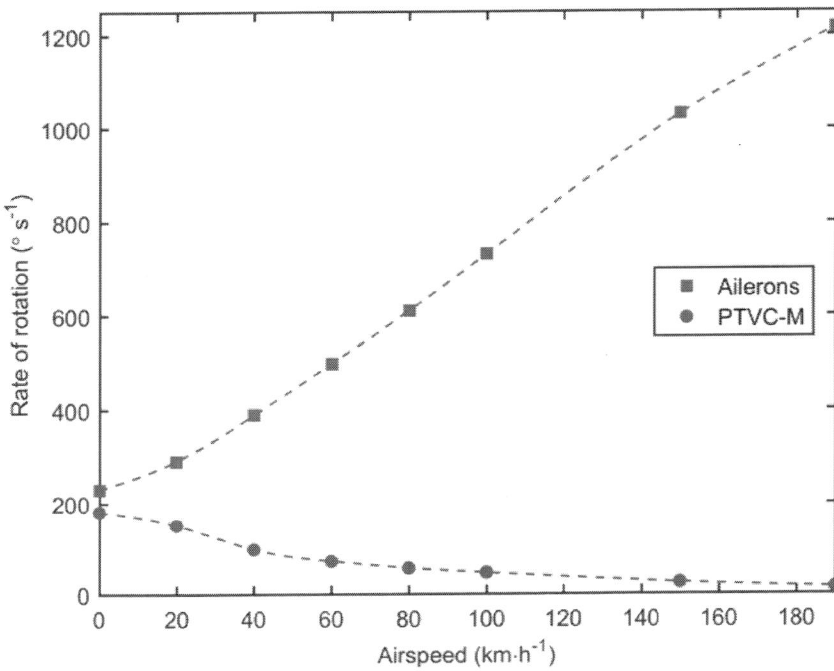

Figure 7. Equilibrium roll rates attributed to ailerons and PTVC-M as a function of airspeed.

The simulated tests were carried out for horizontal wind speeds of 0 km · $h^{-1}$and approximately 40 km · h^{-1} (altitude and terrain dependent) and the results for both scenarios are available in Video 3. The arrows in the second part of the video indicate the direction of wind flow. Also, the local wind intensity and updraft are displayed on the NavGuides. The steady-state stall-spin was generally achieved by applying maximum deflection of rudder and ailerons in the same direction and applying up elevator with no throttle input. For the 0 km · h^{-1} wind, it achieved an equilibrium descent rate of 6.2 ms^{-1}, and the variation between the wind speed and ground speed was ±3 km · h^{-1}, which was taken as the measurement uncertainty of the wind speed following [30] . The spinsonde performance was further characterized at wind intensity of around 190 km · h^{-1} as shown in Figure 8, and the spinsonde continued to work very well despite the uncertainty has slightly increased (with largest deviation being ±8 km · h^{-1}). The aircraft experienced orographic downdrafts because it has drifted behind the hills. The binocular gadget window provides a magnified view of the aircraft in real-time. Note that the ground speed would be equivalent to the speed measurement acquired by the onboard GPS if it were in real field testing [30] .

Figure 8. Evaluating the spinsonde performance in strong wind.

Figure 9. The aircraft performing a powered stall-spin with positive climb rate of 3.5 ms^{-1} at a height of 3 m above the ground. The direction of spin is anticlockwise (viewed from above).

In mesoscale meteorology, the parameter of particular interest to forecasters and the public is the maximum sustained 10-m surface wind [32] , but so far it proved challenging to acquire such measurement using spinsonde technique without incurring risks to the safety of the aircraft [30] . Video 3 shows that PTVC-M not only helped to provide vertical retardation to the spinsonde, but with some throttle input, the aircraft was able to climb while gyrating in the stall-spin. The maneuver was repeated under windy condition with success (Video 3). Figure 9 is a screenshot of the aircraft performing a steady climb at the conclusion of a stall-spin at a height of just 3 meters above ground level with a climb rate of 3.5 ms^{-1}. In other words, the PTVC-M has given the aircraft the ability to climb in a powered stall-spin, serving as a mean to safely exit the maneuver at low altitude, thus making it possible to incorporate the 10-m surface wind into the multi-cycle measurement of vertical wind speed profile.

4. CONCLUSION

We had proposed and demonstrated via simulations the concept of PTVC-M in an attempt to develop ultra-agile VTOL capabilities without relying on large control surfaces which often degrade high-speed flight performance. The small UAV in this work equipped with PTVC-M had demonstrated the ability to perform inverted vertical hover with full 3-axis control. PTVC-M had also enabled the UAV to climb in a stall-spin and such feature will lead to the development of high-performance spinsonde that can directly measure the 10-m surface wind intensity as part of the multi-cycle measurement of the vertical wind speed profile. We believe the proposed PTVC-M will be another indispensable tool in mesoscale meteorology as well as being an enabling technology that brings exciting and enthralling "4D" maneuvers to the world of sport aerobatics.

REFERENCES

1. Wikipedia (2016) Yakovlev Yak-141.

2. Sanchez, A., Escareño, J., Garcia, O. and Lozano, R. (2008) Autonomous Hovering of a Noncyclic Tiltrotor UAV: Modeling, Control and Implementation. Proceedings of the 17th World Congress of the International Federation of Automatic Control, Seoul.

3. Poh, C.H. and Poh, C.K. (2014) Can the Vertical Motions in the Eyewall of Tropical Cyclones Support Persistent UAV Flight? Journal of Aeronautics & Aerospace Engineering, 3, 133.

4. Vargas-Clara, A. and Redkar, S. (2012) Dynamics and Control of a Stop Rotor Unmanned Aerial Vehicle. International Journal of Electrical and Computer Engineering, 2, 597.

5. Australian Aviation (2013) Australian Company Designs New Stoprotor Aircraft Technology. http://australianaviation.com.au/2013/04/australian-company-designs-new-stoprotor-aircraft-technology/

6. ELYTRON Aircraft Inc. (2013) Design. http://elytron.aero/#page-aircraft

7. Latitude Engineering (2016) Hybrid Quadrotor™ Technology.https://latitudeengineering.com/products/hq/

8. ComQuest Ventures LLC (2016) The Vertex VTOL UAS.http://www.comquestventures.com/vertex-vtol-uas/

9. Selig, M.S. (2010) Modeling Full-Envelope Aerodynamics of Small Uavs in Real Time. Proceedings of the AIAA Atmospheric Flight Mechanics Conference, 7635.

10. International Miniature Aerobatic Club (2009) About IMAC. http://www.mini-iac.com/

11. The World Air Sports Federation (2016) F3M Large Radio Control Aerobatics World Cup. http://www.fai.org/world-cups/f3m

12. SEBART International Srl (2016) Sebart 50E ARF Assembly Manual.http://www.sebart.it/download/manuale%20Sebart%2050/Manual%20introduction%20Sebart-50E.pdf

13. Franklin, J.L., Black, M.L. and Valde, K. (2003) GPS Dropwindsonde Wind Profiles in Hurricanes and Their Operational Implications. Weather and Forecasting, 18, 32-44.http://dx.doi.org/10.1175/1520-0434(2003)018<0032:GDWPIH>2.0.CO;2

14. Wikipedia (2016) Dropsonde. http://en.wikipedia.org/wiki/Dropsonde

15. Poh, C.-K. and Poh, C.-H. (2015) Concept of Spinsonde for Multi-Cycle Measurement of Vertical Wind Profile of Tropical Cyclones. Open Journal of Applied Sciences, 5, 145-150.

16. YouTube (2016) Waterfall. https://www.youtube.com/watch?v=feuyIIVRR2w

17. Commercial Aviation Safety Team (2016) Asymmetric Flight. http://www.cast-safety.org/pdf/5_asymmetric_flight.pdf

18. Hepperle, M. (2004) MH 22. http://www.mh-aerotools.de/airfoils/mh22koo.htm

19. Knife Edge Software (2014) KEmax Content Creation Toolkit.http://www.knifeedge.com/KEmax/

20. Great Planes® Model Mfg. (2015) RealFlight® Radio Control Flight Simulator.http://www.realflight.com/

21. Autodesk Inc. (2016) 3ds Max® 3D Modeling, Animation, and Rendering Software.http://www.autodesk.com/products/3ds-max/overview

22. Selig, M.S. (2010) Modeling Propeller Aerodynamics and Slipstream Effects on Small UAVs in Realtime. Proceedings of the AIAA Atmospheric Flight Mechanics Conference, Toronto, 2-5 August 2010, 7938.

23. Armattan Quads (2016) Tricopters. http://www.armattanquads.com/tricopters/

24. Leishman, J.G. (2000) Principles of Helicopter Aerodynamics. 2nd Edition, Cambridge University Press, New York.

25. ArduPilot Development Team (2016) What Is a MultiCopter and How Does It Work?http://ardupilot.org/copter/docs/what-is-a-multicopter-and-how-does-it-work.html

26. Wikipedia (2016) Thrust Vectoring. https://en.wikipedia.org/wiki/Thrust_vectoring

27. Wikipedia (2016) Fairey Rotodyne. https://en.wikipedia.org/wiki/Fairey_Rotodyne

28. Century of Flight (2016) The History of Flight. http://www.century-of-flight.net/Aviation%20history/evolution%20of%20technology/VTOL%20and%20STOL%20Technology%20and%20Aircraft.htm

29. Sport Copter International (2016) About Gyroplanes.http://www.sportcopter.com/Gyroplanes/tabid/124/Default.aspx

30. Hitchens, F. (2015) The Encyclopedia of Aerodynamics. Andrews UK Limited, Bedfordshire.

31. Marchand, A. (2016) A Spanish Pioneer.http://www.airbushelicopters.com/w1/jrotor/71/delacierva.html

32. Gyroplanepassion.com (2016) Inventor of the Autogiro.http://www.gyroplanepassion.com/Juan_de_la_Cierva.html

4
CHAPTER

Advancement of Compositional and Microstructural Design of Intermetallic γ-TiAl Based Alloys Determined by Atom Probe Tomography

Thomas Klein *, Helmut Clemens and Svea Mayer

Department of Physical Metallurgy and Materials Testing, Montanuniversität Leoben, Roseggerstr. 12, 8700 Leoben, Austria

ABSTRACT

Advanced intermetallic alloys based on the γ-TiAl phase have become widely regarded as most promising candidates to replace heavier Ni-base superalloys as materials for high-temperature structural components, due to their facilitating properties of high creep and oxidation resistance in combination with a low density. Particularly, recently developed alloying concepts based on a β-solidification pathway, such as the so-called TNM alloy, which are already incorporated in aircraft engines, have emerged offering the advantage of being processible using near-conventional methods and the option to attain balanced mechanical properties via subsequent heat-treatment. Development trends for the improvement of alloying concepts, especially dealing with issues regarding alloying element distribution, nano-scale phase characterization, phase stability, and phase formation mechanisms demand the utilization of high-resolution techniques, mainly due to the multi-phase nature of advanced TiAl alloys. Atom probe tomography (APT) offers unique possibilities of characterizing chemical compositions with a high spatial resolution and has, therefore, been widely used in recent years with the aim of understanding the materials constitution and appearing basic phenomena on the atomic scale and applying these findings to alloy development. This review, thus, aims at summarizing scientific works regarding the application of atom probe tomography towards the understanding and further development of intermetallic TiAl alloys.

Keywords: titanium aluminides based on γ-TiAl; atom probe tomography; alloy design; microstructure formation; interfacial segregation; phase stability; hardening effects; site-specific specimen preparation

1. INTRODUCTION

Intermetallic titanium aluminides belong to the most promising materials to meet today's most prevalent demands for structural high-temperature materials of combining high strength with low density (next to other important characteristics like high creep and oxidation resistance as well as high modulus and strength retention at elevated temperatures) [1,2,3]. Due to these properties, TiAl alloys are particularly suitable for rotating components of modern propulsion systems, such as turbine blades of aero engines and turbocharger wheels of automotive engines. Especially, tight emission reduction targets, triggered by an increasing global ecological awareness [4], as directed by national and international initiatives and legislation [5,6], have become a major driving force in the development of this alloy class. Incorporation of innovative lightweight components into latest generation propulsion systems results in substantial reductions of greenhouse gases, CO_2 and NO_x, as well as fuel consumption. Consequently, extensive fundamental research projects in close collaboration with industrial partners have been launched to create the basis for knowledge-based compositional and microstructural design and concomitant feasible manufacturing routes. Today, titanium aluminides have become an established material class and sophisticated processing routes are available reflected by an increasing market penetration of TiAl components in both aerospace and automotive industry. Application examples are cast turbine blades of the low-pressure turbine introduced in the GEnx™ by General Electrics [7] and forged turbine blades used in the low-pressure turbine of the recently inaugurated Geared Turbofan™ engine by Pratt and Whitney [8]. For details of current progress in alloy design, processing technologies, applications and prospects, the reader is referred to references [1,2,9,10,11,12].

TiAl alloys consist at room and service temperature entirely of ordered, intermetallic phases, predominantly of γ-TiAl ($L1_0$ structure), α_2-Ti_3Al ($D0_{19}$ structure) and β_o-TiAl (B2 structure). Mainly the inherent brittleness—a result of the intermetallic character of all constituent phases—has made processing a challenging task on the path to industrialization [13]. The combination of thermo-mechanical processing and multiple heat-treatments, which exploit the occurrence of several elapsing phase transformations, has therefore been extensively investigated with the result of enabling the adjustment of different types of microstructures [14]. These in turn allow for tuning of mechanical properties over wide ranges toward the prerequisites of the particular area of operation mainly by the control of morphological parameters, e.g., grain size, colony size and aspect ratio or lamellar interface spacing. In this respect the fully lamellar or nearly fully lamellar microstructures evidence the best combination of strength, creep resistance, ductility, and fracture toughness—all of which are required for feasible manufacturing processes, to achieve balanced mechanical properties as well as to provide structural integrity during a components lifetime [2,14].

In recent years, a novel subclass of so-called β-solidifying TiAl alloys, the TNM alloy, has emerged [2,12,15,16,17]. This alloy solidifies entirely via the disordered β phase (A2 structure) (L → L + β → β ...), which results in an homogeneous, fine grained and almost texture-free microstructure in contrast to peritectically solidifying alloys (L → L + β → α ...), which are prone to segregation, coarse microstructure and a strong casting texture [2,12,15]. TNM alloys comprise, next to Ti and Al, the alloying elements Nb, Mo and small amounts of B, and were designed with the aim of providing a forgeable alloy exhibiting balanced mechanical properties that can be attained via near-conventional manufacturing routes and subsequent heat-treatments [12,18,19]. At elevated temperatures this alloy exhibits a sufficient amount of disordered β phase due to the β-stabilizing elements Nb and Mo, which facilitates hot-workability. This phase is, however, undesirable at service temperatures, but can be removed or reduced via ensuing heat-treatments [12,18]. Figure 1 depicts the microstructure of a TNM alloy in the cast and hot-isostatically pressed (HIP) condition taken by scanning electron microscopy (SEM) in back-scattered electron (BSE) mode. In this micrograph, all major constituent phases are visible and are labeled accordingly. The microstructure is homogeneous, i.e., no significant segregation of alloying elements occurred during processing. Boron, which is added to TNM alloys as a grain refining agent, acts as heterogeneous nucleant during solidification and subsequent phase transformations [20]. Thereby, B is mostly present in thermally stable borides—one of which is visible at the bottom of Figure 1. The depicted cast/HIP microstructure corresponds to the starting condition for following forging operations or to adjust the microstructure directly via refined heat-treatment strategies [21]. In order to enhance the high-temperature capability of TNM alloys, alloying with microalloying elements, such as C and Si, is a suitable means [22,23,24,25]. These alloys are then denoted as TNM^+ alloys and represent a most promising alloy subclass, which effectively increases the maximum operating temperature of TNM alloys.

A minor, but nevertheless important phase, which can hardly be visualized by SEM, due to its extreme fineness, is the ω_o phase, which can occur in many TiAl systems including the TNM alloy [26,27]. In these β_o phase containing TiAl alloys, ω_o particles are observed to precipitate from the $\beta_o$phase and a simultaneous hardening has been reported [28,29,30,31]. However, details of the occurring precipitation reaction are matter of ongoing scientific issues among scientists. As a simultaneous significant solute redistribution is typically observed, atom probe tomography (APT) corresponds to the method of choice for a detailed investigation [22].

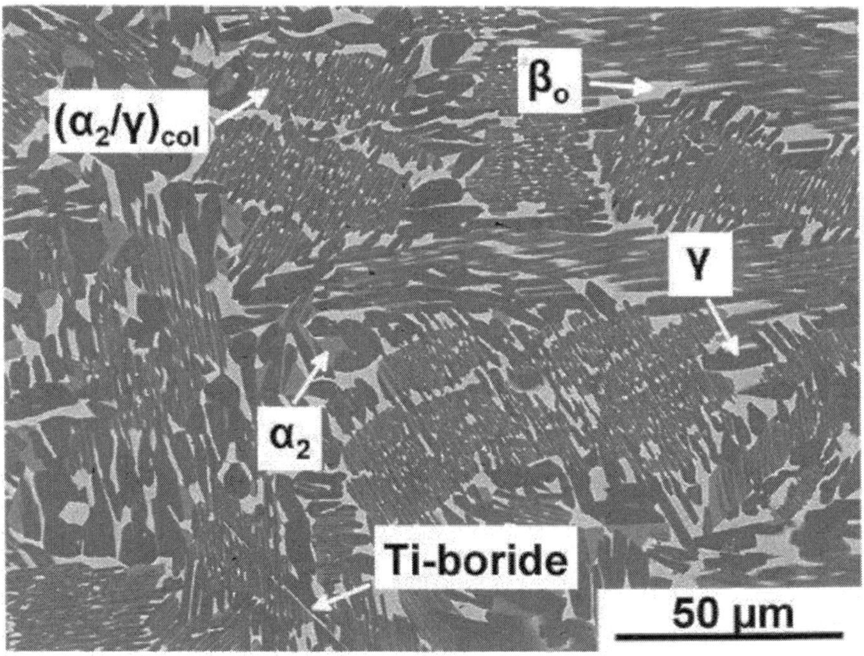

Figure 1. Cast and hot-isostatically pressed (cast/HIP) microstructure of a TNM alloy taken by scanning electron microscopy (SEM) in back-scattered electron (BSE) mode. The microstructure comprises of γ, α_2, and β_o phase of globular morphology as well as of α_2/γ colonies with β_o precipitates in between (secondary precipitates). The γ phase appears dark gray, the α_2 phase is light gray and the β_o phase shows the brightest contrast. Moreover, a rod-shaped Ti-boride is visible at the bottom of this image.

APT (and atom probe field ion microscopy (APFIM)) was extensively applied to many fundamental questions of this alloy class. Two particularities of TiAl alloys account for the intense utilization of APT: (i) mechanical properties, especially ductility, are dependent on the interstitial content of the individual phases [32]; and (ii) as microstructures often consist of extremely fine-scale constituents, high-resolution techniques are required for a detailed characterization [33,34]. The following section will give a brief literature survey encompassing APT studies of the past three decades in the field of titanium aluminides.

Early studies revealed superior properties of α_2/γ two-phase alloys, especially of lamellar structures, particularly in terms of higher strength levels, ductility and fracture toughness in comparison to single-phase alloys, which are either based on γ or α_2. Hence, fundamental research aimed at understanding underlying phenomena. These studies revealed that O impurity levels are highest in the α_2 phase of two-phase alloys [35,36,37]. Its removal from the γ phase was argued to promote ductility by a so-called "scavenging effect", i.e., a purification effect keeping the γ phase ductile [35,36], or entirely by structural features of the lamellar arrangement [37]. Menand et al. [38] rebutted the "scavenging effect" of the $\alpha_2$phase as they were able to evidence that the γ phase of single-phase and two-phase alloys shows equivalent O concentrations, corresponding to the maximum solubility of this phase. As this solubility level is much lower than the alloy's impurity contamination, it is argued that even in high purity single-phase γ-TiAl alloys, the O excess leads to fine-scaled oxide precipitates or a local enrichment, which in turn result in embrittlement. Furthermore, Lefebvre et al. [39] evidenced a strong dependence of the maximum interstitial solubility on the off-stoichiometric composition of the γ phase. It is argued that interstitials such as C, N or O prefer to occupy Ti_6 octahedral sites, which are inherently present in the α_2 phase but not in the γ phase. However, if the phase composition of the γ phase deviates toward the Al-lean side of the stoichiometric composition, the surplus of Ti atoms occupies Al sites, which in turn results in the creation of Ti_6 octahedral cavities. This conclusion was later picked up by Scheu et al. [40] to explain the observed increase in C solubility by alloying with Nb. As Nb preferentially substitutes for Ti in the γ phase [41,42], a surplus of Ti is generated, which locates on Al sites, hence, acts equivalently to off-stoichiometry, effectively increasing the concentration of Ti_6 octahedral sites.

APT was moreover utilized for detailed studies of different alloying concepts mainly in terms of element phase preference and its effect on phase equilibria. TiAl alloys of second and third generation (see references [1,2]) typically contain one or more different transition metal additions. In this respect, Nb, Mo, Ta, W, V, Cr, and Mn have been widely investigated. APT studies of Nb, which is included in most engineering alloys [43,44,45,46,47], revealed only weak partitioning and a homogeneous distribution for both two-phase alloys [48,49,50,51] and three-phase alloys [22,52]. At the same time Mo strongly favors the β_o phase in three-phase alloys [22,52]. In this respect Vb elements generally show a weak phase preference for the β_o phase [22,48,49,50,51,52], whereas VIb elements generally show a strong phase preference for the β_o phase [22,29,52]. The tendency of phase preference corresponds to an increased β-stabilizing effect of VIb elements in comparison to Vb elements [53].

Some alloying elements, moreover, have been shown to effectively increase creep resistance or, more generally, increase the microstructural stability. These effects have been argued to stem from interfacial segregation essentially decreasing interface mobility [54,55]. APT, which is also especially applicable for the study of interfacial segregation due to its superior spatial resolution and the potential of directly quantifying the extent of segregation, evidenced interfacial excesses at phase interfaces of both heavy elements such as W [51,56] and Hf [51] as well as light elements such as C [22].

Progress of recent years in APT technique allows for the generation of reliable data with a reasonable experimental effort. Hence, more sophisticated questions may be answered that arise in current multi-phase multi-component TiAl alloys to improve the existing and to achieve knowledge required for new innovations. Issues of advancing titanium aluminides using APT are characterization of the lamellar structure [51,52], alloying effects [22,40], characterization of ceramic-type strengthening particles, and occurring phase interfaces [57,58] or underlying phenomena of in-service oxygen embrittlement [59], just to name of few examples. This article reviews questions in the development of TiAl alloys that were successfully answered using APT and highlights possibilities of this technique. In particular fundamental questions as well as issues pertaining to the enhancement of TNM alloys are contemplated in the frame of complimentary methods.

2. ATOM PROBE TOMOGRAPHY

The basis of APT has been established in several textbooks and reviews [60,61,62]. Nevertheless, this section should give a short overview of the fundamentals of this technique, the progress that resulted in the state-of-the-art methods as well as the preparation techniques suitable for the specimen production of TiAl alloys.

2.1. Background of Technique: Principles of Method and Compositional Measurements

APT is based on the field evaporation of surface atoms near the tip of a needle-shaped specimen, whereby both time-of-flight mass spectroscopy and position sensitive ion detection are utilized [61,62]. Specimens need to be needle-shaped with tip radii of ≈20–100 nm in order to reach sufficient local electric fields of ≈10–60 V/nm resulting in ionization and emission of surface atoms at cryogenic temperatures and ultra-high vacuum. An approximation of the electric field occurring in the vicinity of an approximately spherical tip is given by Equation (1):

$$F = \frac{V}{k_f \cdot R} \tag{1}$$

where F is the electric field strength induced at the tip's apex with a radius of curvature R and k_f denotes the field factor, which includes the tip shape as well as the electrostatic environment, i.e., corresponds to a geometrical correction stemming from specimen and mounting geometry and V corresponds to the voltage applied [61,62]. From Equation (1), one can recognize the requirement of the specimen's tip to be extremely sharp due to naturally given upper limits of voltage. During the evaporation process, atoms that are positioned close to steps are more likely to evaporate resulting in a self-regulation of the specimen geometry. The evaporated ions follow the electric field lines created between the specimen and the detector. The evaporation process is controlled concomitantly to the measure of the time of departure by creation of a standing field below the evaporation field of the atoms of the material investigated. This field is then superimposed either by voltage pulses or laser (thermal) pulses—both of which result in a controlled emission of ions. These short duration pulses allow for a distinct discrimination of the time-of-flight of each ion that hits the position sensitive detector. The information gained is the mass-to-charge ratio for each evaporated ion. Data post-processing such as background subtraction or peak deconvolution is often required to identify the chemical identity of a particular species and, thereby, create an accurate chemical analysis. Moreover, the three-dimensional specimen volume can be reconstructed using the information determined by the position sensitive detector providing knowledge of the spatial distribution of the atoms in, e.g., phases or clusters in a very descriptive way [61].

2.2. The Local Electrode Atom Probe

The introduction of the local electrode atom probe (LEAP™) [63,64] has allowed for a widened utilization of APT and much work has been done since. Recent advances of method and data evaluation are explained in detail in references [64,65,66,67,68,69]. In the LEAP™ instrument a standing voltage is applied between the specimen and the counter electrode, which is, as in the case of any atom probe, superimposed by high voltage or laser pulses. The novelty is the introduction of the local electrode, whereby the specimen is positioned very close to this electrode as schematically depicted in Figure 2. This experimental arrangement allows for a dramatic increase of measureable sample volumes and an extended field of view. Evaporated ions travel through a hole in the local electrode toward the position sensitive detection system, whereby the generated signal as well as its evaluation and interpretation may be performed equivalently as described in Section 2.1.

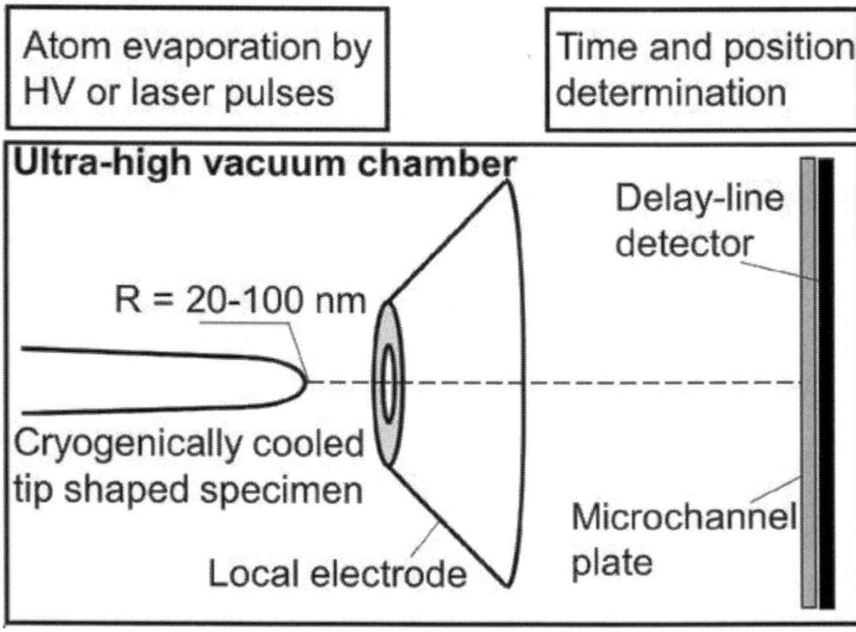

Figure 2. Scheme of the major components present in a local electrode atom probe (LEAP) redrawn after reference [62]. Surface atoms near the tip's apex are evaporated either by high voltage or laser pulses and accelerated by standing voltage toward the position sensitive detector. Time-of-flight mass spectroscopy allows for identification of the chemical identity; position determination enables a three-dimensional reconstruction of the measured specimen volume.

Further enhancement of mass resolution may be attained via utilization of a reflectron belonging to the so-called techniques of energy compensation or flight time compensation. To this end, the travelling ions enter an electrostatic field and are, thereby, deflected onto a curved pathway. Ions of lower energy will be deflected first and, hence, travel a shorter path than ions of higher energy. A longer path increases the time-of-flight and vice versa, i.e., the peak width in the time-of-flight spectrum is narrowed corresponding to an increased mass resolution potential. However, one major drawback of the reflectron system is the concomitant reduction of detection efficiency.

2.3. Methods of Specimen Preparation

There exist two distinct preparation routes that are suitable and necessary for the study of TiAl alloys by means of APT. The first approach follows a sequence that is schematically depicted in Figure 3. Thereby, blanks possessing ≈1–2 cm length and a quadratic cross-section of ≈300 μm side length are usually cut using precision cutting tools. These are then mounted into Cu (or Ni) crimps and are etched in a first step (Figure 3, Step 1) using a combination of electrolyte and inert liquid to create a neck roughly in the middle of the specimen. Subsequently, the inert layer is removed and the specimen is etched until it fractures (Step 2). Following this procedure, final shaping and sharpening is conducted using a microloop with a drop of electrolyte (Step 3). For TiAl alloys an etchant consisting of 5% perchloric acid in acetic acid has been successfully used for Steps 1 and 2 using polishing voltages of ≈25 V and ≈15 V, respectively. The best results of final tip shaping were obtained using 2% perchloric acid in butoxyethanol at ≈10 V. For further details on electropolishing routines and other preparation methods the reader is referred to the textbooks [60,61].

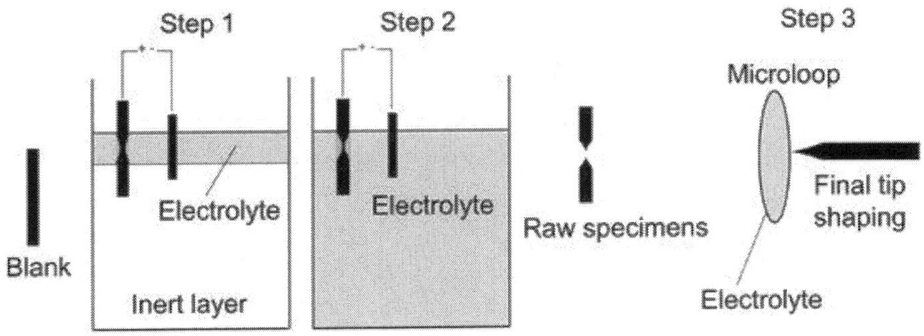

Figure 3. Scheme of the specimen preparation routine by electropolishing. The sample material is cut into prismatic blanks and mounted into Cu crimps. Subsequently, these blanks are etched and sharpened using three steps, the first two resulting in a raw tip-shaped geometry, which has, however, to be further sharpened using a microloop procedure denoted as Step 3. Drawn after references [60,61].

APT specimens can be prepared alternatively using the focused ion beam (FIB) based lift-out technique as described in detail in references [70,71,72,73,74,75]. This procedure is usually carried out in a dual beam FIB device equipped with both an electron as well as an ion source. The sequence of steps conducted is represented schematically in Figure 4. In a first step the SEM is simply used to identify a region of interest for the following measurement. Subsequently, the material is protected from Ga implantation and concomitant sample contamination and amorphization by a Pt deposit. A lamella is then formed upon excavation at two sides contiguous to the Pt deposit using the ion beam. The lamella can then be lifted out after cutting free using a micromanipulator and can be mounted to any kind of pre-tip. This can be the posts of a Si needle pad, tips generated on a cut and electropolished transmission electron microscopy (TEM) grid [73,75,76] or a single tip. At last the final tip shape is generated using annular milling, whereby voltage and current are gradually reduced to decrease Ga contamination [77].

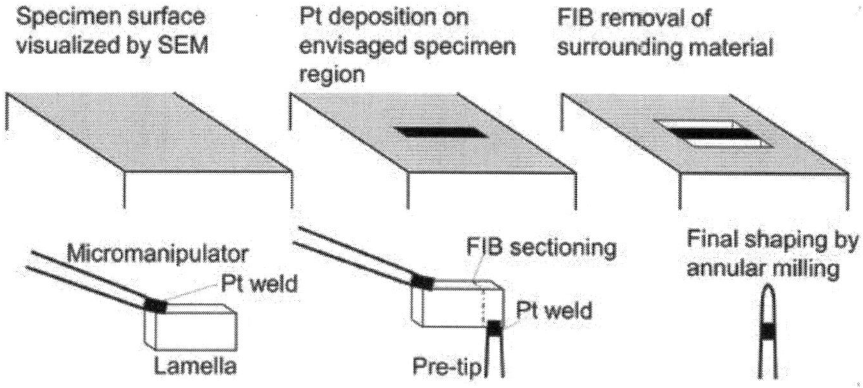

Figure 4. Scheme of the atom probe tomography (APT) specimen preparation by lift-out technique. In the first step a region of interest is identified using SEM and subsequently protected from Ga implantation by a Pt deposit. Contiguous material is then excavated at both sides generating a lamella, which can then be removed using a micromanipulator. Following this sequence, parts of the lamella can be transferred to any kind of suitable pre-tip. Once mounted, the final tip shape can be obtained by annular milling. Schematically redrawn according to references [70,71].

3. EFFECTS OF ALLOYING ELEMENTS

3.1. Niobium and Molybdenum: Phase Formation, Phase Preference and Phase Stability

This section will focus on the effects of the decisive substitutional alloying elements that are typically used in advanced TiAl alloys as studied by APT and will view them in the context of established alloying related effects. Especially, different transition metals have been identified as valuable additions to increase both ductility and strength. For example V, Cr, Mn, Nb, Mo, Ta, and W are all among most common additions to TiAl alloys [15,17,53,78]. However, effects of Nb and Mo are of particular importance for the understanding of the TNM alloy and related alloys as they constitute the major alloying elements [79,80,81].

Niobium has been identified as a crucial element for the enhancement of high-temperature strength [47,82,83] and oxidation resistance [43]. The strengthening effect observed in two-phase alloys (comprising of α_2 and γ phase) has been recognized to stem mainly from an increasing amount of $\alpha_2$and from an occurring structural refinement [84]. In the case of β-solidifying alloys, β_o phase is retained to room temperature, which is due to the fact that Nb acts as a β-stabilizing element. Hence, the Nb distribution in the constituent phases is expected to differ from classic two-phase TiAl alloys and is, thus, of particular interest as strong partitioning potentially results in microsegregation, i.e., in microstructural inhomogeneities, that could deteriorate creep and fatigue resistance [83]. Therefore, the local chemical composition of the individual phases of a TNM alloy in the homogenized condition was determined by APT, as reported in reference [22]. Chemical compositions are depicted in Figure 5 in the form of a bar chart, neglecting for the moment the occurrence of a minor phase fraction of ω_o phase, which will be dealt with in Section 3.3. Considering Ti and Al, it is of particular interest to note that the composition of the phases α_2 and β_o deviates significantly from the compositions of the ideal binary phases (Ti_3Al and TiAl, respectively). This strong deviation of both phases is due to differing site occupation behavior of the transition metal atoms that are alloyed in the TNM system. While all transition metals prefer to substitute at Ti sites in the α_2 phase, there is a strong trend for the substitution at Al sites in the β_o phase, as suggested by ab initio calculations in reference [85]. Niobium shows a similar distribution between the individual phases with a weak preference for the β_o phase only. This observation is in agreement to the fact that Nb was identified as a β-stabilizing element [2,53]. However, although its addition results in the introduction of the β_o phase, which is stable at room and service temperature, there are still significant amounts dissolved in the α_2 and γ phase, i.e., the enrichment within the β_o phase does not entirely remove Nb from the other phases. In contrast, Mo strongly partitions between the phases, whereby the β_o phase is most favored. This observation corresponds well to the fact that Mo has been reported to be a by far stronger β-stabilizing element than Nb [2,53]. This effect has far-reaching implications on processing, e.g., on the hot-workability. Since additions of Mo, already in modest amounts, stabilizes the disordered β phase at elevated temperatures, hot-working becomes possible using near conventional hot-working equipment [18,19]. Most importantly, the volume fraction of β/β_o phase can be reduced significantly on ensuing heat-treatment as its presence deteriorates the creep resistance [86,87]. This is only possible due to the particular curve progression of phase fractions occurring in the TNM system, i.e., the course of the β phase fraction shows a pronounced C-shape with a minimum between 1250 °C and 1280 °C [12,88]. Using a second heat-treatment step, which is typically conducted slightly above the envisaged service temperature, the amount of β phase can be further reduced due to the fact that the stabilization of the β phase by β-stabilizing elements diminishes in the course of the ordering reaction, i.e., Mo strongly stabilizes the disordered β phase, but hardly stabilizes its ordered counterpart β_o [80]. Moreover, a fine lamellar structure is adjusted and the phase fractions are equilibrated, yielding a reduced contribution of phase transformation to the creep elongation [2,18].

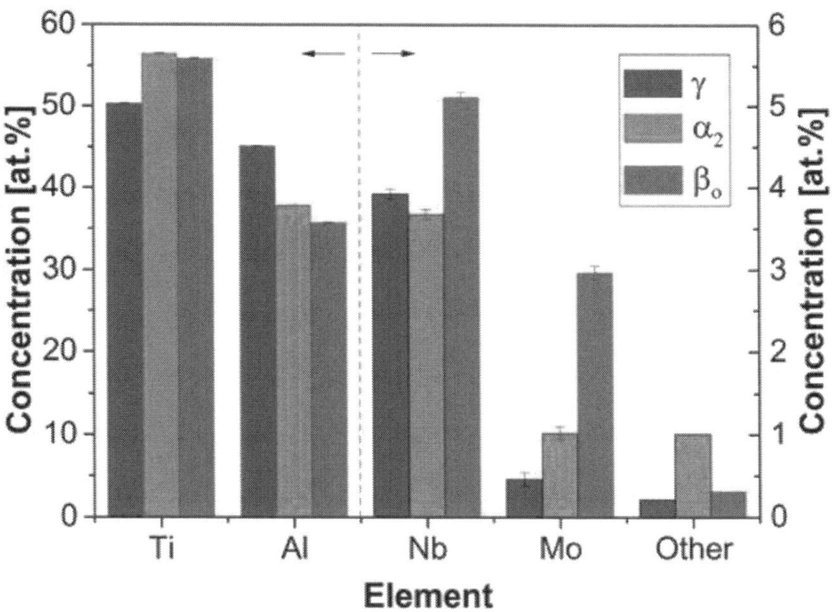

Figure 5. Phase composition of a TNM alloy in the homogenized condition. Concentrations of the major alloying elements, Ti and Al, are displayed on the left ordinate, whereas concentrations of the minor alloying elements, Nb, Mo as well as impurities denoted as "Other" are given on the right ordinate. Titanium shows a slightly higher concentration in α_2 and β_o than in the γ phase and at the same time the Al concentration is reduced.

This enormous difference in phase composition in comparison to the ideal binary phases (Ti_3Al and TiAl, respectively) is due to preferred lattice site occupation of transition metal atoms on Ti sites in the α_2 phase and Al sites in the β_o phase. Nb and Mo both show a tendency to accumulate in the β_o phase corresponding to their well-established β-stabilizing effect, which is more pronounced in case of Mo. Impurities, mainly constituted by interstitial elements like C, N or O (here their sum is referred to as "Other"), prefer to locate in the $\alpha_2$phase.

Impurity levels are most pronounced in the α_2 phase as discernible from Figure 5, where the sum of all impurities present in this alloy is shown. This observation is due to the fact that most impurities correspond to interstitial elements, which are mostly dissolved in the α_2 phase due to the inherent presence of Ti_6 octahedral sites, where interstitials are incorporated favorably. Both other phases do not constitute any of these sites intrinsically. Some other impurities, however, corresponding to Fe, Cr and V, as identified by APT, all of which represent typical β-stabilizing elements, contribute to the increased impurity levels present in the β_o phase in comparison to the γ phase.

3.2. Boron Addition and Boride Formation

Boron has been identified as a particularly valuable alloying element in TiAl alloys [20,89]. Early studies revealed a significant grain refinement effect, which was attributed to the presence of thermally stable borides that act as heterogeneous nucleation sites during solidification and on ensuing phase transformation [89]. Some details about the grain refinement effects, especially, as they depend on the evolving phase transformation sequence, i.e., refining effects differ from β-solidifying alloys to peritectic alloys and the cooling rate applied, remain, however, matter of debate and ongoing research [90]. A detailed analysis of the chemical composition of these borides can help to understand, e.g., effects of alloying elements on their stability. Larson et al. [91] determined chemical compositions of different Ti-borides occurring in a Ti-47Al-2Cr-2Nb-0.15B (in atom%, at%) (TIA-20) alloy using APFIM. Results of a measured mono-boride are reproduced in Table 1 and are compared to the chemical composition we have determined for a boride present in the TNM alloy as visible in Figure 1. In our case, the investigated specimen was prepared by site-specific specimen preparation using the FIB technique. Apparently, according to the evaluated chemical composition, this boride similarly corresponds to a mono-boride. Interestingly, the boride contained in the TNM alloy exhibits a significant and by far greater amount of Nb than in the TIA-20 alloy and also some Mo. The boride observed in the TNM alloy, however, contains ≈45 at% B only, i.e., apparently its composition is off-stoichiometric either by creation of anti-site defects or vacancies possibly due to the alloyed elements. The presented results, moreover, evidence that Ti-borides occurring in TiAl alloys show hardly any solubility for Al and are comprised mainly of Ti and B.

Table 1. Boride composition determined as reported in reference [91] and in the present work. The standard error is given next to each concentration. Both particles correspond to mono-borides in accordance with a MB stoichiometry, where M represents the constituting metal atoms.

Alloy	Element Concentration (at%)					
	B	Ti	Al	Nb	Mo	Other
TIA-20 [91] [1]	52.0 ± 3.2	46.3 ± 3.2	0.4 ± 0.4	1.2 ± 0.7	-	0.1
TNM (present study)	45.1 ± 0.01	47.0 ± 0.01	-	6.8 ± 0.03	1.1 ± 0.04	0.0

[1] Actual alloy composition: Ti-47Al-2Cr-2Nb-0.15B (at%).

3.3. The Formation of ω_o Phase

As mentioned previously most engineering TiAl alloys that are currently in use are alloyed with modest to high amounts of Nb [46,82], which is generally vital for high-temperature strength as well as creep and oxidation resistance. The introduction of this element, however, results in an increased susceptibility to the formation of additional phases such as the ω_o phase [26,27,28,29,31], which was observed to result in an increasing hardness [22,28]. In the case of β-solidifying TiAl alloys, this phase precipitates from the β_o phase at temperatures below ≈825 °C [28] in contrast to alloys free of β_o phase, where precipitation from the α_2 phase can occur [92,93]. For the detailed understanding of the precipitation process occurring in TNM alloys, it is of particular interest to understand how the elements redistribute between β_o and ω_o phase during the precipitation sequence and, thereby, withdraw information pertaining to stabilization and destabilization. Hence, APT specimens of a heat-treated TNM sample containing β_o and ω_o phase were prepared site-specifically as described in reference [75]. There, a TNM specimen that was annealed above the ω_o-solvus temperature and cooled slowly in the furnace, allowing for the precipitation reaction to take place, as reported in reference [22]. The presence of both β_0 and ω_0 phase was verified by complimentary transmission Kikuchi diffraction conducted directly at the site-specifically prepared APT [22,75]. An APT reconstruction containing the alloying elements Nb (green line) and Mo (red line) is shown in Figure 6a, where isoconcentration surfaces at 1.3 at% Mo are also indicated. A clearly inhomogeneous distribution of these elements is discernable and the phase decomposition occurring can be visualized explicitly by the isoconcentration surfaces. As these surfaces are created with the aid of a concentration gradient it should be noted that they do not necessarily coincide the actual crystallographic phase boundaries. Some of the ω_0 particles show a parallel arrangement elongated along preferred growth directions that correspond to the established orientation relationship (OR) $\langle 111\rangle_{\beta o}\{11\bar{0}\}_{\beta o} \| [0001]_{\omega o}\{11\bar{2}0\}_{\omega o}$ [94].

For a quantification of the changes in chemical composition across the interfaces of Figure 6a, proximity histograms of the major alloying elements Ti, Al, Nb, and Mo were calculated and are depicted in Figure 6b. The phase boundary appears to be chemically diffuse, while being crystallographically sharp. The Al distribution seems nearly homogeneous, whereas Nb and Mo show opposing partitioning in comparison to Ti. The Mo redistribution occurring during growth of the $\omega_0$particles is clearly the most pronounced effect and is indicative for an actual phase decomposition. Apparently, Mo is rejected from the ω_0 phase and enriched in the β_0 matrix in agreement with references [22,28]. In the vicinity of the interface, a Mo pile-up occurs as visible in Figure 6b. The appearing diffusional redistribution is presumed to limit the formation kinetics.

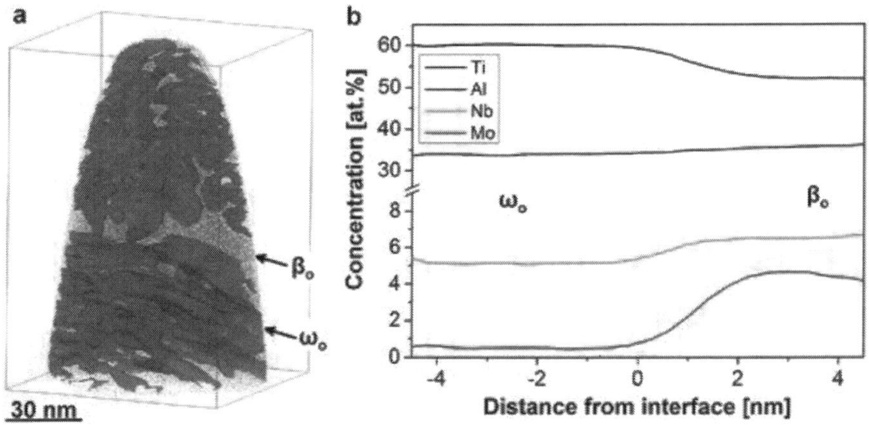

Figure 6. APT reconstruction containing β_o and ω_o phase and proximity histogram. (**a**) Only Nb (green) and Mo atoms (red) are indicated for clarity. Moreover, isoconcentration surfaces have been generated at 1.3 at% Mo. These interfaces clearly separate regions of β_o and ω_o phase. ω_o precipitates show parallel arrangements, which correspond to the predominating orientation relationship (OR) (see text); (**b**) Concentration profiles calculated as proximity histograms across the interfaces depicted in (**a**) are shown. The concentration of Al remains nearly unaltered, whereas Mo and Nb on the one hand and Ti on the other hand show an opposing partitioning behavior.

3.4. Enhancement of the TNM Alloying Concept: Effects of Carbon Addition

The enhancement of high-temperature strength has constantly been a goal in the design of novel TiAl alloys. In this regard particularly alloying with the element C has been investigated. This interstitial element strengthens the material either by solid solution hardening or by carbide precipitation [32,95,96]. The predominating mechanism is, thereby, determined by the alloy composition and the materials' thermal history. Carbon is mainly located in the direct environment of six Ti atoms, i.e., in so-called Ti_6 octahedral cavities, as suggested in references [38,39]. Following this train of thought, the $\alpha_2$phase is argued to show the highest intrinsic solubility as it is the only phase in the compositional range of engineering TiAl alloys that intrinsically possesses this type of cavities, whereas the both other phases, γ and β_o, do not possess any. In case of the γ phase a strong dependence of the maximum solubility for C is observed with respect to the Al concentration. In particular, toward the Al-lean off-stoichiometric composition, the C solubility increases [39]. This observation is justified by the argument that the excess of Ti atoms tend to occupy Al sites and, thereby, introduce Ti_6 octahedral cavities that can be occupied by C. Similarly, the addition of alloying elements, especially transition metal additions such as Nb, increases the solubility limit for interstitials in the γ phase [40]. The mechanism is contended to be related to the fact that Nb preferentially occupies Ti sites [41,42] in the $L1_0$ lattice and, thus, similarly as in the case of off-stoichiometry introduces a surplus of Ti atoms, which relocate to the Al sites, whereby also similarly favorable cavities are created. In case of the β_o phase there are no favorable sites present in the stoichiometric B2 structure, which can, however, not be introduced by substitution, as all transition metal atoms preferentially occupy Al sites [85]. Therefore, the β_o phase is expected to show a very limited C solubility. In order to verify these theoretical considerations, the chemical composition of the individual phases of a model TNM alloy that was additionally alloyed with 0.75 at% C (TNM0.75C) was determined by APT in the homogenized condition after forging as reported in references [22,97]. In agreement with the theory on C distribution, the C concentrations were found to follow the sequence $cC\alpha2 > cC\gamma > cC\beta o$ as indicated in Figure 7.

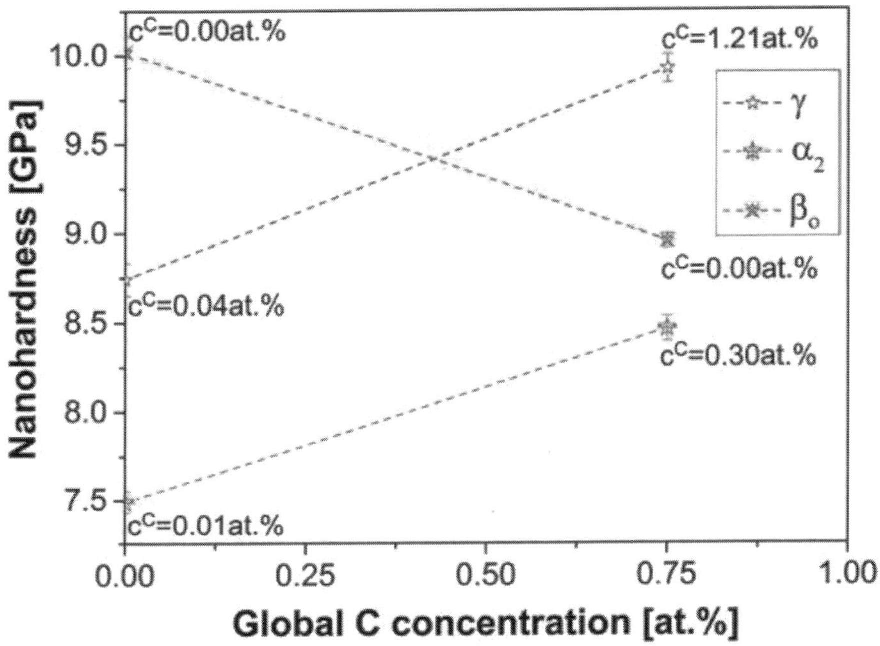

Figure 7. Hardness as measured by nanoindentation of the individual phases of a TNM alloy and a C-containing derivative, TNM0.75C. The error bars indicated correspond to the standard errors of the mean. In the C-free state the β_o phase is the hardest phase followed by α_2 and γ. By alloying with C, the latter two phases harden via solid solution hardening, whereas the β_o phase softens due to the absence of ω_o particles. The C concentration of each individual phase of both alloys as measured by APT is labeled next to the respective hardness values. Dashed lines are indicated only to guide the eye.

Moreover, in this figure the local mechanical properties of the constituting phases as determined by nanoindentation are correlated with the global C concentration [22].

Nanoindentation is, in this respect, a particularly suitable method as it allows withdrawing information about mechanical properties of individual microstructural constituents, which can be correlated to the local chemical compositions analyzed by APT. Apparently, α_2 and γ show an increase in nano-hardness by the incorporation of C, which can be explained by a solid solution hardening mechanism as no carbides were observed in the investigated material condition. This effect is in agreement to the C concentrations determined by APT. The nano-hardness of the β_o phase, however, decreases with increasing global C content, which is interesting to note, as the β_o phase does not show any solubility for C. It has been evidenced in the literature that C is a strong stabilizer of the α_2 phase, while the phase fraction of β_o is reduced by its incorporation [23]. As mentioned before, Mo tends to partition strongly to the β_o phase, i.e., if the $\beta_o$phase fraction is reduced, the amount of Mo in this phase increases. An increasing amount of Mo tends to inhibit the formation of ω_o phase as discussed in Section 3.3. The absence of ω_o particles in the interior of the β_o phase of the C-containing alloy, results in a reduced nano-hardness as no precipitation hardening contributes [28].

Research on C-containing TiAl alloys has, moreover, evidenced the significant impact of C on the microstructural stability in terms of resistance against discontinuous precipitation of lamellar structures [24] as well as reduced recrystallization kinetics after hot-working [98]. These effects, on the one hand, are due to the impaired redistribution of the strongly partitioning element C required for the formation of new grains during any kind of recrystallization. On the other hand, solute-drag effects may be responsible for a reduced interfacial mobility [99]. In order to visualize and quantify the presence and extent of such segregation effects, APT specimens were prepared site-specifically using a dual beam FIB as reported in

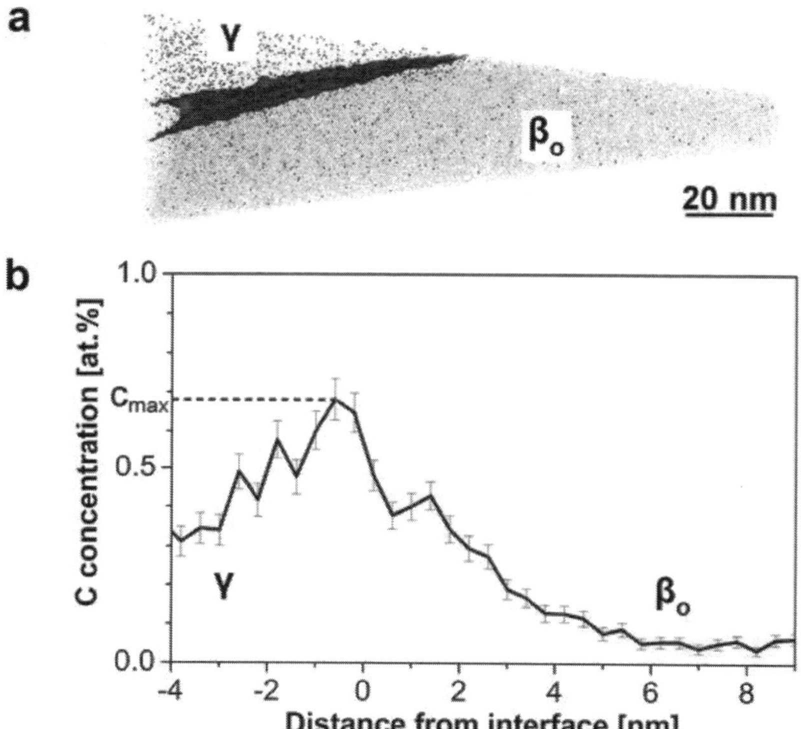

Figure 8. (**a**) APT reconstruction of a TNM0.75C specimen containing a β_o/γ interface; and (**b**) C proximity histogram across this interface. (**a**) A clearly inhomogeneous distribution of the alloying elements Mo (red) and C (blue) across the interface is visible. The interface indicated corresponds to a Mo isoconcentration surface, which shows the most pronounced concentration gradient. The histogram in (**b**) evidences significant segregation of C to the β_o/γ interface, with a local maximum C concentration of ≈0.7 at%. (Reprinted with permission from [22], Copyright© 2015, Elsevier).

references [22,75]. A reconstruction and a proximity histogram of a TNM0.75C specimen are represented in Figure 8. In this case the tip contains a β_o/γ interface and only the minor alloying elements Mo (in red) and C (in blue) are depicted. Obviously, an inhomogeneous distribution is evident in Figure 8a corresponding to prevailing strong partitioning. The proximity histogram calculated across this interface demonstrates quantitatively significant local C enrichment in the vicinity of the interface with a local maximum C concentration of ≈0.7 at%. This enrichment exemplifies clearly that interfacial segregation of C is existent, which in turn increases microstructural stability of TNM alloys, when alloyed with C [23,24].

4. CHARACTERIZATION OF NANO-SCALED LAMELLAR STRUCTURES: EFFECTS OF SILICON ADDITION

A further crucial alloying element that is often added to TiAl alloys is Si. Although its effect is controversially discussed in the literature, the general consensus points toward an increase in creep resistance via stabilization of the microstructure [100,101,102]. In particular, the precipitation of ζ-silicides (Ti_5Si_3) at lamellar interfaces is observed frequently in many alloy systems, which is argued to stabilize these interfaces as well as to reduce dislocation mobility. The specific precipitation mechanism, which ensues especially in two-phase alloys with a low α_2 volume fraction due to heat-treatment and creep, leads to aligned particles at lamellar interfaces [103]. The α_2 phase contains the most Si, while the solubility for Si is rather limited in the γ phase. Hence, silicides precipitate once the α_2 phase fraction reaches a lower margin, which restricts also the global solubility for Si. However, advanced engineering TiAl alloys, such as the TNM alloy, typically contain higher amounts of α_2 phase and, furthermore, a varying amount of β_o phase—both of which may alter to evolving reaction sequence. Therefore, a nano-lamellar structure was generated via a two-step heat-treatment in a TNM alloy containing Si as reported in reference [52]. During the first heat-treatment step, the solution heat-treatment conducted above the γ-solvus temperature, only disordered α and β phases are present [88,104]. On subsequent rapid cooling lamellar γ phase formation is suppressed, which can be finely and homogeneously precipitated on ensuing annealing as demonstrated in references [105,106,107]. The annealing procedure is, thereby, typically conducted slightly above the envisaged service temperature. During this stage γ lamellae form in the supersaturated α_2 grains upon sufficient thermal activation by the creation of two Shockley partial dislocations bordering a stacking fault that locally accomplishes the change in crystal structure [108,109,110]. Figure 9 depicts such a structure generated in a TNM^+ alloy (Ti-43.5Al-4Nb-1Mo-0.1B-0.3C-0.3Si in at%) after heat-treatment at 1340 °C for 15 min followed by oil quenching and subsequent annealing at 800 °C followed by furnace cooling. In Figure 9a a SEM micrograph is shown, but the ultra-fine lamellar structure within the α_2/γ colonies cannot be resolved. Between these colonies the prior $\beta_o$phase is visible, which decomposed during annealing into γ platelets, ω_o particles and remaining $\beta_o$phase. The lamellar structure can be resolved by TEM as depicted in Figure 9b. The interfaces are clearly delineated and appear smooth without significant interfacial roughness. All γ lamellae of a prior $\alpha_2$grain are aligned parallel due to the Blackburn OR [108]. Details regarding the mechanism of lamellar structure formation can be found in references [33,109,110]. Most importantly, no ζ-type precipitates are observed in the vicinity of the lamellar interfaces.

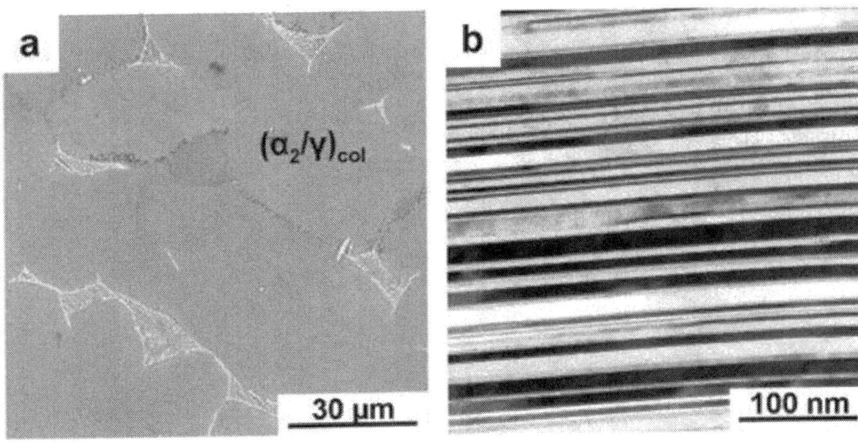

Figure 9. Microstructure after two-step heat-treatment taken by (**a**) SEM in BSE mode and (**b**) transmission electron miscroscopy (TEM) bright field image, as reported in reference [52]. Prior supersaturated α_2 grains have decomposed to α_2/γ colonies, which cannot be resolved using SEM. Lamellae of each grain show only one orientation variant according to the Blackburn OR. Between lamellar colonies β_o phase and several transformation products are visible (see text). The nano-lamellar structure can be evidenced by TEM (**b**), where the specimen has been tilted into edge-on condition.

APT reconstructions of the individual alloying elements corresponding to the lamellar structure depicted in Figure 9 are shown in Figure 10. In each image, the lamellar structure is apparent to a greater or lesser extent, corresponding to their partitioning between the α_2 and γ lamellae. The partitioning of

C and Si toward the α_2 phase is particularly distinct with Si showing a more pronounced tendency to partition to this phase, which is in agreement with reported literature data [111,112] (seeSection 3.4 for a more detailed discussion of the accumulation of C in the α_2 phase). Although these elements show an isovalent electronic structure, the dominating mechanism must be different as C is dissolved interstitially, while Si is dissolved substitutionally as suggested in references [52,85] by ab initio calculations.

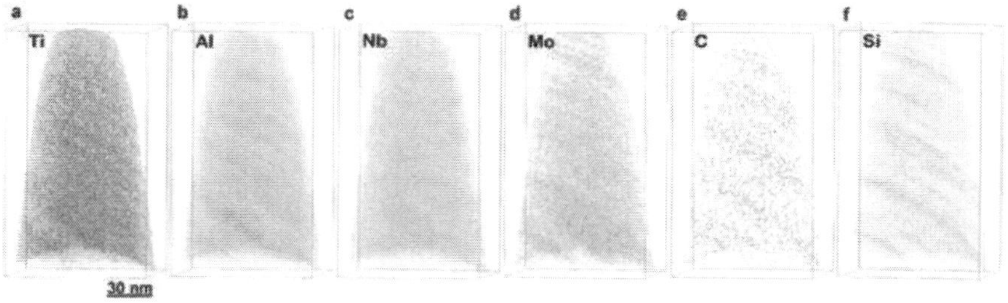

Figure 10. Reconstructions of the lamellar structure of a TNM$^+$ alloy. In all images, the lamellar structure is visible to different extents corresponding to different partitioning ratios. Carbon shows a tendency to accumulate in the α_2 phase and Si partitions even more pronounced to this phase. See Table 2 for quantitative analysis.

The chemical composition of the lamellar structure depicted in Figure 10 was quantified (Table 2) after separation of the adjacent phases by isoconcentration surfaces defined at a Ti value of 55 at%. Subsequently, the compositional evaluation was conducted by peak deconvolution, which is required due to overlaps in the mass spectrum by different chemical species due to the multi-component nature of the TiAl alloy investigated. The chemical composition of the γ phase is comparably close to its equilibrium composition. However, a deficiency of Al is observed, which is interesting to note as Nb and Mo preferentially substitute for Ti [41,85]. It is, however, energetically favorable to substitute for Ti, which in turn relocates to Al sites as recently shown by ab initio calculations [85]. In case of the α_2 phase the off-stoichiometry is even more pronounced, i.e., the Ti concentration is reduced and the Al concentration is increased. Apparently, the surplus of Al occupies Ti sites and equivalently do the alloyed transition metal atoms in agreement with the data presented in Figure 5 and reference [85]. Nb shows in the lamellar structure a weak preference for the γ phase, which has similarly been reported for the lamellar structure of two-phase alloys [51]. The opposite is occurring in case of Mo. As mentioned before both C and Si prefer to locate in the α_2 phase. In case of C this is due to the fact that this phase intrinsically possesses Ti_6 octahedral sites. Apparently, in case of Si the chemical environment and crystal structure of α_2 results in an energetically preferred environment, where it preferentially substitutes for Al [85]. These chemical details demonstrate the origin of the absence of ζ-silicides in TNM$^+$ alloys in the material condition investigated as suggested by Klein et al. [52]. The α_2 phase is capable of storing substantial amounts of Si and, therefore, dominates the global solubility limit. TNM alloys contain substantial amounts of α_2 phase due to lower amounts of Al in comparison to classic TiAl alloys. Hence, all Si present in the interior of each α_2/γ colony is dissolved and no interfacial precipitation occurs.

Table 2. Phase composition of the lamellar structure depicted inFigure 10 as studied in reference [52]. The errors given correspond to the standard errors. Nb shows a weak preference for the γ phase, whereas Mo evidences a weak preference for the α_2 phase. C and Si prefer to locate in the α_2 phase.

Phase	Element Concentration (at%)						
	Ti	Al	Nb	Mo	C	Si	Other
γ	49.8 ± 0.02	44.8 ± 0.02	4.01 ± 0.03	0.85 ± 0.03	0.16 ± 0.03	0.21 ± 0.02	0.17
α_2	61.4 ± 0.03	32.1 ± 0.05	3.90 ± 0.05	1.14 ± 0.06	0.35 ± 0.06	0.51 ± 0.06	0.60

5. CONCLUSIONS

Intermetallic materials based on γ-TiAl have become an engineering material class in the last years and are now increasingly incorporated into novel propulsion concepts in jet and automotive engines. Especially β-solidifying alloys, such as the TNM alloy, have shown to be processible using robust production routes and to achieve reasonable mechanical properties. APT has played a pivotal role in the understanding and the in-depth analysis of these alloys, which has been summarized in this work. APT has been shown to be a powerful characterization tool shedding light upon chemical phase composition, chemical fluctuations, i.e., early stages of precipitation by compositional clustering, partitioning behavior of elements as well as interface characteristics. These phenomena can often solely be analyzed when APT is employed. Thereby, valuable information for the understanding and enhancement of γ-TiAl based alloys is generated. In this work, major findings using APT in correlation with other techniques have been reviewed in the context of existing literature. The major conclusions can be summarized as follows:

- The chemical composition of all constituent phases (γ, α_2, β_o) of a TNM alloy was determined. Preferential partitioning of Nb and Mo to the β_o phase is discernable and the α_2 phase shows the highest impurity levels. The element B is mainly present in mono-borides.

- Within the β_o phase the precipitation of a further minor phase, the so-called ω_o phase, can be detected, which is depleted in the strong β-stabilizing element Mo.

- Carbon, if alloyed to TNM alloys, preferentially resides in the α_2 phase, followed by the γ phase. The β_o phase appears to be completely depleted of C. In the microstructural condition investigated both α_2 and γ phase are significantly strengthened by solid solution hardening due to the presence of the interstitial element C. However, the β_o phase seems to soften by the addition of C, which stems from the fact that the presence of C yields a lower β_o phase fraction, thereby increasing the Mo content in this phase, which in turn results in the suppression of the ω_o phase formation and its concomitant hardening effects. Carbon, moreover, tends to segregate to interfaces, which may result in lower interface mobilities, thus, stabilizing the microstructure.

- In ultra-fine lamellar structures, Si preferentially occupies the α_2 phase, whereby neither interfacial segregation nor interfacial precipitation is observed in TNM^+ alloys, in contrast to other TiAl alloying systems, which is due to a substantial fraction of α_2 phase where Si is substitutionally dissolved.

ACKNOWLEDGMENTS

Parts of the research presented in this review were conducted in the framework of the FFG project "energy drive", Research Studios Austria, Austria, grant agreement No. 832040.

AUTHOR CONTRIBUTIONS

All authors contributed equally to this work. T.K. wrote the bulk of the manuscript in collaboration with the other authors. The contribution of the authors to the experimental work is specified in the respective original works that are cited throughout this publication.

CONFLICTS OF INTEREST

The authors declare no conflict of interest.

REFERENCES

1. Appel, F.; Paul, J.D.H.; Oehring, M. *Gamma Titanium Aluminide Alloys. Science and Technology*; Wiley-VCH: Weinheim, Germany, 2011.

2. Clemens, H.; Mayer, S. Design, Processing, Microstructure, Properties, and Applications of Advanced Intermetallic TiAl Alloys. *Adv. Eng. Mater.* **2013**, *15*, 191–215.

3. Wu, X. Review of alloy and process development of TiAl alloys. *Intermetallics* **2006**, *14*, 1114–1122.

4. Lee, D.S.; Pitari, G.; Grewe, V.; Gierens, K.; Penner, J.E.; Petzold, A.; Prather, M.J.; Schumann, U.; Bais, A.; Berntsen, T. Transport impacts on atmosphere and climate: Aviation. *Atmos. Environ.* **2010**, *44*, 4678–4734.

5. ACARE Advisory Council for Aviation Research and Innovation in Europe. Available online: http://www.acare4europe.org/ (accessed on 18 April 2016).

6. European Commission Plans Legislative Framework to Ensure the EU Meets Its Target for Cutting CO_2 Emissions from Cars. European Comission Web Site. Available online: http://europa.eu/rapid/press-release_IP-07-155_en.htm (accessed on 18 April 2016).

7. Bewlay, B.P.; Weimer, M.; Kelly, T.; Suzuki, A.; Subramanian, P.R. The Science, Technology, and Implementation of TiAl Alloys in Commercial Aircraft Engines. In *Intermetallic-Based Alloys—Science, Technology and Applications*; Baker, I., Heilmaier, M., Kumar, S., Yoshimi, K., Eds.; Cambridge University Press: Cambridge, UK, 2013; Volume 1516, pp. 49–58.

8. Habel, U.; Heutling, F.; Helm, D.; Kunze, C.; Smarsly, W.; Das, G.; Clemens, H. Forged Intermetallic γ-TiAl Based Alloy Low Pressure Turbine Blade in the Geared Turbofan. In Proceedings of the 13th World Conference on Titanium, San Diego, CA, USA, 16–20 August 2015; Venkatesh, V., Pilchak, A.L., Allison, J.E., Ankem, S., Boyer, R., Christodoulou, J., Fraser, H.L., Ashraf Imam, M., Kosaka, Y., Rack, H.J., et al., Eds.; John Wiley & Sons, Inc.: Hoboken, NJ, USA, 2016; pp. 1223–1227.

9. Clemens, H.; Smarsly, W.; Güther, V.; Mayer, S. Advanced Intermetallic Titanium Aluminides. In Proceedings of the 13th World Conference on Titanium, San Diego, CA, USA, 16–20 August 2015; Venkatesh, V., Pilchak, A.L., Allison, J.E., Ankem, S., Boyer, R., Christodoulou, J., Fraser, H.L., Imam, M.A., Kosaka, Y., Rack, H.J., et al., Eds.; John Wiley & Sons, Inc.: Hoboken, NJ, USA, 2016; pp. 1189–1200.

10. Clemens, H.; Mayer, S. Intermetallic titanium aluminides in aerospace applications—Processing, microstructure and properties. *Mater. High Temp.* **2016**, *33*, 1–11.

11. Clemens, H.; Mayer, S. Intermetallic Titanium Aluminides as Innovative High Temperature Lightweight Structural Materials—How Materialographic Methods Have Contributed to Their Development. *Pract. Metallogr.* **2015**, *52*, 691–721.

12. Clemens, H.; Wallgram, W.; Kremmer, S.; Güther, V.; Otto, A.; Bartels, A. Design of Novel β-Solidifying TiAl Alloys with Adjustable β/B2-Phase Fraction and Excellent Hot-Workability. *Adv. Eng. Mater.* **2008**, *10*, 707–713.

13. Lasalmonie, A. Intermetallics: Why is it so difficult to introduce them in gas turbine engines?*Intermetallics* **2006**, *14*, 1123–1129.

14. Kim, Y.-W. Effects of microstructure on the deformation and fracture of γ-TiAl alloys. *Mater. Sci. Eng. A* **1995**, *192–193*, 519–533.

15. Imayev, R.M.; Imayev, V.M.; Oehring, M.; Appel, F. Alloy design concepts for refined gamma titanium aluminide based alloys. *Intermetallics* **2007**, *15*, 451–460.

16. Tetsui, T.; Shindo, K.; Kobayashi, S.; Takeyama, M. A newly developed hot worked TiAl alloy for blades and structural components. *Scr. Mater.* **2002**, *47*, 399–403.

17. Jin, Y.; Wang, J.N.; Yang, J.; Wang, Y. Microstructure refinement of cast TiAl alloys by β solidification. *Scr. Mater.* **2004**, *51*, 113–117.

18. Wallgram, W.; Schmölzer, T.; Cha, L.; Das, G.; Güther, V.; Clemens, H. Technology and mechanical properties of advanced γ-TiAl based alloys. *Int. J. Mater. Res.* **2009**, *100*, 1021–1030.

19. Huber, D.; Werner, R.; Clemens, H.; Stockinger, M. Influence of process parameter variation during thermo-mechanical processing of an intermetallic β-stabilized γ-TiAl based alloy. *Mater. Charact.* **2015**, *109*, 116–121.

20. Hecht, U.; Witusiewicz, V.; Drevermann, A.; Zollinger, J. Grain refinement by low boron additions in niobium-rich TiAl-based alloys. *Intermetallics* **2008**, *16*, 969–978.

21. Schwaighofer, E.; Clemens, H.; Mayer, S.; Lindemann, J.; Klose, J.; Smarsly, W.; Güther, V. Microstructural design and mechanical properties of a cast and heat-treated intermetallic multi-phase γ-TiAl based alloy. *Intermetallics* **2014**, *44*, 128–140.

22. Klein, T.; Schachermayer, M.; Mendez-Martin, F.; Schöberl, T.; Rashkova, B.; Clemens, H.; Mayer, S. Carbon distribution in multi-phase γ-TiAl based alloys and its influence on mechanical properties and phase formation. *Acta Mater.* **2015**, *94*, 205–213.

23. Schwaighofer, E.; Rashkova, B.; Clemens, H.; Stark, A.; Mayer, S. Effect of carbon addition on solidification behavior, phase evolution and creep properties of an intermetallic β-stabilized γ-TiAl based alloy. *Intermetallics* **2014**, *46*, 173–184.

24. Kastenhuber, M.; Rashkova, B.; Clemens, H.; Mayer, S. Enhancement of creep properties and microstructural stability of intermetallic β-solidifying γ-TiAl based alloys. *Intermetallics* **2015**, *63*, 19–26.

25. Schwaighofer, E.; Clemens, H.; Lindemann, J.; Stark, A.; Mayer, S. Hot-working behavior of an advanced intermetallic multi-phase γ-TiAl based alloy. *Mater. Sci. Eng. A* **2014**, *614*, 297–310.

26. Bendersky, L.A.; Boettinger, W.J.; Burton, B.P.; Biancaniello, F.S.; Shoemaker, C.B. The formation of ordered ω-related phases in alloys of composition Ti_4Al_3Nb. *Acta Metall. Mater.* **1990**, *38*, 931–943.

27. Strychor, R.; Williams, J.C.; Soffa, W.A. Phase transformations and modulated microstructures in Ti-Al-Nb alloys. *Metall. Trans. A* **1988**, *19*, 225–234.

28. Schloffer, M.; Rashkova, B.; Schöberl, T.; Schwaighofer, E.; Zhang, Z.; Clemens, H.; Mayer, S. Evolution of the ω_o phase in a β-stabilized multi-phase TiAl alloy and its effect on hardness. *Acta Mater.* **2014**, *64*, 241–252.

29. Song, L.; Xu, X.J.; You, L.; Liang, Y.F.; Lin, J.P. Phase transformation and decomposition mechanisms of the $\beta_o(\omega)$ phase in cast high Nb containing TiAl alloy. *J. Alloys Compd.* **2014**, *616*, 483–491.

30. Song, L.; Xu, X.; You, L.; Liang, Y.; Lin, J. Ordered ω phase transformations in Ti-45Al-8.5Nb-0.2B alloy. *Intermetallics* **2015**, *65*, 22–28.

31. Stark, A.; Bartels, A.; Clemens, H.; Schimansky, F.-P. On the Formation of Ordered ω-phase in High Nb Containing γ-TiAl Based Alloys. *Adv. Eng. Mater.* **2008**, *10*, 929–934.

32. Perdrix, F.; Trichet, M.-F.; Bonnentien, J.-L.; Cornet, M.; Bigot, J. Relationships between interstitial content, microstructure and mechanical properties in fully lamellar Ti-48Al alloys, with special reference to carbon. *Intermetallics* **2001**, *9*, 807–815.

33. Denquin, A.; Naka, S. Phase transformation mechanisms involved in two-phase TiAl-based alloys—I. Lamellar structure formation. *Acta Mater.* **1996**, *44*, 343–352.

34. Appel, F.; Wagner, R. Microstructure and deformation of two-phase γ-titanium aluminides. *Mater. Sci. Eng. R* **1998**, *22*, 187–268.

35. Denquin, A.; Naka, S.; Huguet, A.; Menand, A. Atom-probe investigation of the partitioning of interstitial elements in two-phase $\gamma+\alpha_2$ TiAl-based alloys. *Scr. Metall. Mater.* **1993**, *28*, 1131–1136.

36. Uemori, R.; Hanamura, T.; Morikawa, H. Oxygen scavenging effect of the α_2 phase in the TiAl intermetallic compound. *Scr. Metall. Mater.* **1992**, *26*, 969–974.

37. Huguet, A.; Menand, A. Atom-probe determination of interstitial element concentration in two-phase and

single-phase TiAl-based alloys. *Appl. Surf. Sci.* **1994**, *76–77*, 191–197.

38. Menand, A.; Huguet, A.; Nérac-Partaix, A. Interstitial solubility in γ and α_2 phases of TiAl-based alloys. *Acta Mater.* **1996**, *44*, 4729–4737.

39. Lefebvre, W.; Menand, A.; Loiseau, A.; Blavette, D. Atom probe study of phase transformations in a Ti–48 at.% Al alloy. *Mater. Sci. Eng. A* **2002**, *327*, 40–46.

40. Scheu, C.; Stergar, E.; Schober, M.; Cha, L.; Clemens, H.; Bartels, A.; Schimansky, F.-P.; Cerezo, A. High carbon solubility in a γ-TiAl-based Ti–45Al–5Nb–0.5C alloy and its effect on hardening.*Acta Mater.* **2009**, *57*, 1504–1511.

41. Boll, T.; Al-Kassab, T.; Yuan, Y.; Liu, Z.G. Investigation of the site occupation of atoms in pure and doped TiAl/Ti_3Al intermetallic. *Ultramicroscopy* **2007**, *107*, 796–801.

42. Al-Kassab, T.; Yuan, Y.; Kluthe, C.; Boll, T.; Liu, Z.-G. Investigation of the ordering and atomic site occupancies of Nb-doped TiAl/Ti_3Al intermetallics. *Surf. Interface Anal.* **2007**, *39*, 257–261.

43. Kim, B.G.; Kim, G.M.; Kim, C.J. Oxidation behavior of TiAl-X (X = Cr, V, Si, Mo or Nb) intermetallics at elevated temperature. *Scr. Metall. Mater.* **1995**, *33*, 1117–1125.

44. Lin, J.P.; Zhao, L.L.; Li, G.Y.; Zhang, L.Q.; Song, X.P.; Ye, F.; Chen, G.L. Effect of Nb on oxidation behavior of high Nb containing TiAl alloys. *Intermetallics* **2011**, *19*, 131–136.

45. Pilone, D.; Felli, F.; Brotzu, A. High temperature oxidation behaviour of TiAl–Cr–Nb–Mo alloys.*Intermetallics* **2013**, *43*, 131–137.

46. Paul, J.D.H.; Appel, F.; Wagner, R. The compression behaviour of niobium alloyed γ-titanium alumindies. *Acta Mater.* **1998**, *46*, 1075–1085.

47. Appel, F.; Paul, J.D.H.; Oehring, M.; Clemens, H.; Fischer, F.D. Physical metallurgy of high Nb-containing TiAl alloys. *Z. Metallkunde* **2004**, *95*, 585–591.

48. Qin, G.W.; Smith, G.; Inkson, B.J.; Dunin-Borkowski, R. Distribution behaviour of alloying elements in $\alpha_2(\alpha)/\gamma$ lamellae of TiAl-based alloy. *Intermetallics* **2000**, *8*, 945–951.

49. Larson, D.; Liu, C.; Miller, M. Microstructural characterization of segregation and precipitation in $\alpha_2 + \gamma$ titanium aluminides. *Mater. Sci. Eng. A* **1997**, *239–240*, 220–228.

50. Larson, D.; Liu, C.; Miller, M. The alloying effects of tantalum on the microstructure of an $\alpha_2 + \gamma$ titanium aluminide. *Mater. Sci. Eng. A* **1999**, *270*, 1–8.

51. Gerstl, S.; Kim, Y.-W.; Seidman, D.N. Atomic Scale Chemistry of α_2/γ Interfaces in a Multi-Component TiAl Alloy. *Interface Sci.* **2004**, *12*, 303–310.

52. Klein, T.; Rashkova, B.; Holec, D.; Clemens, H.; Mayer, S. Silicon distribution and silicide precipitation during annealing in an advanced multi-phase γ-TiAl based alloy. *Acta Mater.* **2016**,*110*, 236–245.

53. Kainuma, R.; Fujita, Y.; Mitsui, H.; Ohnuma, I.; Ishida, K. Phase equilibria among α (hcp), β (bcc) and γ ($L1_0$) phases in Ti–Al base ternary alloys. *Intermetallics* **2000**, *8*, 855–867.

54. Karthikeyan, S.; Viswanathan, G.B.; Gouma, P.I.; Vasudevan, V.K.; Kim, Y.-W.; Mills, M.J. Mechanisms and effect of microstructure on creep of TiAl-based alloys. *Mater. Sci. Eng. A* **2002**,*329–331*, 621–630.

55. Karadge, M.; Kim, Y.W.; Gouma, P.I. Synergistic precipitation strengthening in TiAl alloys. *Appl. Phys. Lett.* **2006**, *89*, 181921.

56. Larson, D.J.; Liu, C.T.; Miller, M.K. Tungsten segregation in $\alpha_2 + \gamma$ titanium aluminides.*Intermetallics* **1997**, *5*, 497–500.

57. Gerstl, S.; Seidman, D. Chemical and Structural Investigation of Internal Domains of Needle-Like Ti_3AlC Carbide Precipitates in γ-TiAl with 3-D Atom-Probe Tomography. *Microsc. Microanal.***2006**, *12*, 1570–1571.

58. Gerstl, S.; Seidman, D.; Gribb, A.A.; Kelly, T.F. Leap microscopes look at TiAl alloys. *Adv. Mater. Proc.* **2004**, *162*, 31–33.

59. Draper, S.L.; Isheim, D. Environmental embrittlement of a third generation γ TiAl alloy.*Intermetallics* **2012**, *22*, 77–83.

60. Miller, M.K.; Cerezo, A.; Hetherington, M.G.; Smith, G.*Atom Probe Field Ion Microscopy*; Clarendon Press: Oxford, UK, 1996.

61. Gault, B. *Atom Probe Microscopy*; Springer Series in Materials Science 160; Springer: New York, NY, USA, 2012.

62. Marquis, E.A.; Hyde, J.M. Applications of atom-probe tomography to the characterisation of solute behaviours. *Mater. Sci. Eng. R* **2010**, *69*, 37–62.

63. Kelly, T.F.; Camus, P.P.; Larson, D.J.; Holzman, L.M.; Bajikar, S.S. On the many advantages of local-electrode atom probes. *Ultramicroscopy* **1996**, *62*, 29–42.

64. Kelly, T.F.; Miller, M.K. Invited review article: Atom probe tomography. *Rev. Sci. Instrum.* **2007**, *78*, 31101.

65. Cerezo, A.; Clifton, P.H.; Galtrey, M.J.; Humphreys, C.J.; Kelly, T.F.; Larson, D.J.; Lozano-Perez, S.; Marquis, E.A.; Oliver, R.A.; Sha, G.; et al. Atom probe tomography today. *Mater. Today* **2007**, *10*, 36–42.

66. Miller, M.K.; Forbes, R.G. Atom probe tomography. *Mater. Charact.* **2009**, *60*, 461–469.

67. Miller, M.K.; Kelly, T.F.; Rajan, K.; Ringer, S.P. The future of atom probe tomography. *Mater. Today***2012**, *15*, 158–165.

68. Gault, B.; Moody, M.P.; Cairney, J.M.; Ringer, S.P. Atom probe crystallography. *Mater. Today* **2012**,*15*, 378–386.

69. Cairney, J.M.; Rajan, K.; Haley, D.; Gault, B.; Bagot, P.A.J.; Choi, P.-P.; Felfer, P.J.; Ringer, S.P.; Marceau, R.K.W.; Moody, M.P. Mining information from atom probe data. *Ultramicroscopy* **2015**,*159*, 324–337.

70. Miller, M.K.; Russell, K.F.; Thompson, G.B. Strategies for fabricating atom probe specimens with a dual beam FIB. *Ultramicroscopy* **2005**, *102*, 287–298.

71. Miller, M.K.; Russell, K.F. Atom probe specimen preparation with a dual beam SEM/FIB miller.*Ultramicroscopy* **2007**, *107*, 761–766.

72. Thompson, K.; Lawrence, D.; Larson, D.J.; Olson, J.D.; Kelly, T.F.; Gorman, B. In situ site-specific specimen preparation for atom probe tomography. *Ultramicroscopy* **2007**, *107*, 131–139.

73. Felfer, P.J.; Alam, T.; Ringer, S.P.; Cairney, J.M. A reproducible method for damage-free site-specific preparation of atom probe tips from interfaces. *Microsc. Res. Tech.* **2012**, *75*, 484–491.

74. Babinsky, K.; de Kloe, R.; Clemens, H.; Primig, S. A novel approach for site-specific atom probe specimen preparation by focused ion beam and transmission electron backscatter diffraction.*Ultramicroscopy* **2014**, *144*, 9–18.

75. Schachermayer, M.; Klein, T.; Clemens, H.; Mayer, S. Preparation Methods for Examining the ω_o-Phase Formation in a β-Solidifying TiAl Alloy via Atom Probe Tomography. *Pract. Metallogr.* **2016**,*53*, 73–85.

76. Herbig, M.; Choi, P.; Raabe, D. Combining structural and chemical information at the nanometer scale by correlative transmission electron microscopy and atom probe tomography.*Ultramicroscopy* **2015**, *153*, 32–39.

77. Thompson, K.; Gorman, B.; Larson, D.; van Leer, B.; Hong, L. Minimization of Ga Induced FIB Damage Using Low Energy Clean-up. *Microsc. Microanal.* **2006**, *12*, 1736–1737.

78. Takeyama, M.; Kobayashi, S. Physical metallurgy for wrought gamma titanium aluminides.*Intermetallics* **2005**, *13*, 993–999.

79. Schmoelzer, T.; Mayer, S.; Sailer, C.; Haupt, F.; Güther, V.; Staron, P.; Liss, K.-D.; Clemens, H. In Situ Diffraction Experiments for the Investigation of Phase Fractions and Ordering Temperatures in Ti-44 at% Al-(3-7) at% Mo Alloys. *Adv. Eng. Mater.* **2011**, *13*, 306–311.

80. Holec, D.; Legut, D.; Isaeva, L.; Souvatzis, P.; Clemens, H.; Mayer, S. Interplay between effect of Mo and chemical disorder on the stability of β/β_o-TiAl phase. *Intermetallics* **2015**, *61*, 85–90.

81. Kabra, S.; Yan, K.; Mayer, S.; Schmoelzer, T.; Reid, M.; Dippenaar, R.; Clemens, H.; Liss, K.-D. Phase transition and ordering behavior of ternary Ti–Al–Mo alloys using in-situ neutron diffraction. *Int. J. Mater. Res.* **2011**, *102*, 697–702.

82. Chen, G.; Zhang, W.J.; Liu, Z.C.; Li, S.J.; Kim, Y.-W. Microstructure and properties of high-Nb containing TiAl-base alloys. In *Gamma Titanium Aluminides*; Kim, Y.-W., Dimiduk, D.M., Loretto, M.H., Eds.; The Minerals, Metals & Materials Society: Warrendale, PA, USA, 1999; pp. 371–380.

83. Chen, G.L.; Xu, X.J.; Teng, Z.K.; Wang, Y.L.; Lin, J.P. Microsegregation in high Nb containing TiAl alloy ingots

beyond laboratory scale. *Intermetallics* **2007**, *15*, 625–631.

84. Appel, F.; Oehring, M.; Wagner, R. Novel design concepts for gamma-base titanium aluminide alloys. *Intermetallics* **2000**, *8*, 1283–1312.

85. Holec, D.; Reddy, R.K.; Klein, T.; Clemens, H. Preferential site occupancy of alloying elements in TiAl-based phases. *J. Appl. Phys.* **2016**, *119*, 205104.

86. Appel, F.; Paul, J.D.H.; Oehring, M.; Fröbel, U.; Lorenz, U. Creep behavior of TiAl alloys with enhanced high-temperature capability. *Metall. Mater. Trans. A* **2003**, *34*, 2149–2164.

87. Hamzah, E.; Kanniah, M.; Harun, M. Effect of Beta Phase on Room to High Temperature Mechanical Properties of As-Cast Gamma Titanium Aluminide. *Mech. Adv. Mater. Struct.* **2009**, *16*, 384–389.

88. Schmoelzer, T.; Liss, K.-D.; Zickler, G.A.; Watson, I.J.; Droessler, L.M.; Wallgram, W.; Buslaps, T.; Studer, A.; Clemens, H. Phase fractions, transition and ordering temperatures in TiAl–Nb–Mo alloys: An in- and ex-situ study. *Intermetallics* **2010**, *18*, 1544–1552.

89. Hu, D. Role of boron in TiAl alloy development: A review. *Rare Met.* **2016**, *35*, 1–14.

90. Oehring, M.; Stark, A.; Paul, J.D.H.; Lippmann, T.; Pyczak, F. Microstructural refinement of boron-containing β-solidifying γ-titanium aluminide alloys through heat treatments in the β phase field.*Intermetallics* **2013**, *32*, 12–20.

91. Larson, D.J.; Liu, C.T.; Miller, M.K. Boron solubility and boride compositions in $\alpha_2 + \gamma$ titanium aluminides. *Intermetallics* **1997**, *5*, 411–414.

92. Song, L.; Xu, X.; You, L.; Liang, Y.; Wang, Y.; Lin, J. Ordered α_2 to ω_o phase transformations in high Nb-containing TiAl alloys. *Acta Mater.* **2015**, *91*, 330–339.

93. Bystrzanowski, S.; Bartels, A.; Stark, A.; Gerling, R.; Schimansky, F.-P.; Clemens, H. Evolution of microstructure and texture in Ti–46Al–9Nb sheet material during tensile flow at elevated temperatures. *Intermetallics* **2010**, *18*, 1046–1055.

94. Sikka, S.K.; Vohra, Y.K.; Chidambaram, R. Omega phase in materials. *Prog. Mater. Sci.* **1982**, *27*, 245–310.

95. Worth, B.D.; Jones, W.J.; Allison, J.E. Creep deformation in near-γ TiAl: II. Influence of carbon on creep deformation in Ti-48Al-1V-0.3C. *Metall. Mater. Trans. A* **1995**, *26*, 2961–2972.

96. Tian, W.H.; Nemoto, M. Effect of carbon addition on the microstructures and mechanical properties of γ-TiAl alloys. *Intermetallics* **1997**, *5*, 237–244.

97. Klein, T.; Mendez-Martin, F.; Schachermayer, M.; Rashkova, B.; Clemens, H.; Mayer, S. Distribution of Alloying Elements within the Constituent Phases of a C-containing γ-TiAl Based Alloy studied by Atom Probe Tomography. In *Advanced Structural and Functional Intermetallic-Based Alloys*; Baker, I., Heilmaier, M., Kishida, K., Mills, M., Miura, S., Eds.; Cambridge University Press: Cambridge, UK, 2015; Volume 1760, pp. 7–12.

98. Stark, A.; Bartels, A.; Gerling, R.; Schimansky, F.-P.; Clemens, H. Microstructure and Texture Formation during Hot Rolling of Niobium-Rich γ TiAl Alloys with Different Carbon Contents.*Adv. Eng. Mater.* **2006**, *8*, 1101–1108.

99. Cahn, J.W. The impurity-drag effect in grain boundary motion. *Acta Metall.* **1962**, *10*, 789–798.

100. Tsuyama, S.; Mitao, S.; Minakawa, K.-N. Alloy modification of γ-base titanium aluminide for improved oxidation resistance, creep strength and fracture toughness. *Mater. Sci. Eng. A* **1992**, *153*, 451–456.

101. Noda, T.; Okabe, M.; Isobe, S.; Sayashi, M. Silicide precipitation strengthened TiAl. *Mater. Sci. Eng. A* **1995**, *192–193*, 774–779.

102. Kim, Y.W.; Kim, S.L. Effects of microstructure and C and Si additions on elevated temperature creep and fatigue of gamma TiAl alloys. *Intermetallics* **2014**, *53*, 92–101.

103. Karthikeyan, S.; Mills, M.J. The role of microstructural stability on compression creep of fully lamellar γ-TiAl alloys. *Intermetallics* **2005**, *13*, 985–992.

104. Erdely, P.; Werner, R.; Schwaighofer, E.; Clemens, H.; Mayer, S. In-situ study of the time-temperature-transformation behaviour of a multi-phase intermetallic β-stabilised TiAl alloy.*Intermetallics* **2015**, *57*, 17–24.

105. Cha, L.; Scheu, C.; Clemens, H.; Chladil, H.F.; Dehm, G.; Gerling, R.; Bartels, A. Nanometer-scaled lamellar microstructures in Ti–45Al–7.5Nb–(0; 0.5)C alloys and

their influence on hardness.*Intermetallics* **2008**, *16*, 868–875.

106. Cha, L.; Schmoelzer, T.; Zhang, Z.; Mayer, S.; Clemens, H.; Staron, P.; Dehm, G. In Situ Study of γ-TiAl Lamellae Formation in Supersaturated α_2-Ti_3Al grains. *Adv. Eng. Mater.* **2012**, *14*, 299–303.

107. San Juan, J.; Simas, P.; Schmoelzer, T.; Clemens, H.; Mayer, S.; Nó, M.L. Atomic relaxation processes in an intermetallic Ti–43Al–4Nb–1Mo–0.1B alloy studied by mechanical spectroscopy.*Acta Mater.* **2014**, *65*, 338–350.

108. Blackburn, M.J. Some aspects of phase transformations in titanium alloys. In *The Science, Technology and Application of Titanium*; Jaffee, R.I., Promisel, N.E., Eds.; Pergamom Press: Oxford, UK, 1970.

109. Pond, R.C.; Shang, P.; Cheng, T.T.; Aindow, M. Interfacial dislocation mechanism for diffusional phase transformations exhibiting martensitic crystallography: Formation of TiAl + Ti_3Al lamellae.*Acta Mater.* **2000**, *48*, 1047–1053.

110. Zghal, S.; Thomas, M.; Naka, S.; Finel, A.; Couret, A. Phase transformations in TiAl based alloys.*Acta Mater.* **2005**, *53*, 2653–2664.

111. Hsu, F.-Y.; Klaar, H.-J.; Wang, G.-X.; Dahms, M. Influence of Si content on microstructure of TiAl alloys. *Mater. Charact.* **1996**, *36*, 371–378.

112. Huang, Z.W. Thermal stability of Ti-44Al-4Nb-4Hf-0.2Si-1B alloy. *Intermetallics* **2013**, *37*, 11–21.

5
CHAPTER

Graphene/Epoxy Coating as Multifunctional Material for Aircraft Structures

Tullio Monetta, Annalisa Acquesta * and Francesco Bellucci

Department of Chemical, Materials and Production Engineering, University of Naples Federico II, Piazzale Tecchio 80, 80125 Naples, Italy

ABSTRACT

Recently, the use of graphene as a conductive nanofiller in the preparation of inorganic/polymer nanocomposites has attracted increasing interest in the aerospace field. The reason for this is the possibility of overcoming problems strictly connected to the aircraft structures, such as electrical conductivity and thus lightning strike protection. In addition, graphene is an ideal candidate to enhance the anti-corrosion properties of the resin, since it absorbs most of the light and provides hydrophobicity for repelling water. An important aspect of these multifunctional materials is that all these improvements can be realized even at very low filler loadings in the polymer matrix. In this work, graphene nanoflakes were incorporated into a water-based epoxy resin, and then the hybrid coating was applied to Al 2024-T3 samples. The addition of graphene considerably improved some physical properties of the hybrid coating as demonstrated by Electrochemical Impedance Spectroscopy (EIS) analysis, ameliorating anti-corrosion performances of raw material. DSC measurements and Cross-cut Test showed that graphene did not affect the curing process or the adhesion properties. Moreover, an increment of water contact angle was displayed.

Keywords: multifunctional material; graphene/epoxy coatings; corrosion resistance.

1. INTRODUCTION

Use of graphene as nanofiller in polymer matrices to form advanced multifunctional materials is one of the most promising routes in aerospace, automobile and defence industries [1,2]. The construction of a 3D compactly interconnected graphene network can offer a significant increase in electrical and thermal conductivity of polymer composites. Since a growing and inevitable issue for aircraft structures is corrosion protection, graphene, owing to its barrier properties, could be used to enhance corrosion resistance. The surprising finding is that a very small amount of graphene can significantly improve the physical properties of neat polymers. These outstanding properties of the graphene can be combined with polymer matrix properties to develop multifunctional coatings. In recent years, the number of publications about graphene-based multifunctional materials has increased dramatically [3,4,5,6,7,8,9,10,11,12], but in literature very few papers use these nanocomposites as coatings. In particular, research about improvement of corrosion resistance applying graphene/epoxy coatings are even less common [13,14,15,16]. No paper addresses the possibility of using a water-based epoxy resin, considering the new regulations about the emission of volatile organic compounds [17], and no paper applies the graphene-based coating to the most common of aircraft alloys, that is, Al 2024-T3. Due to the possible multifunctional properties of the graphene/epoxy coating, this study, starting from the evaluation of the barrier properties and adhesion, continues on to an analysis of the electrical conductivity, the UV resistance and the galvanic coupling between aluminium substrate and graphene nanoflakes. The motivation is to examine if the addition of graphene can improve the above physical characteristics of neat polymer. In present paper, results obtained by studies on barrier and adhesion properties are showed.

The graphene/epoxy coating was prepared by incorporating a very small amount of graphene nanoflakes into a water-based epoxy resin, mixing in an ultrasonic bath and applying the hybrid coating to 2024-T3 aluminium alloy panels. The analysis of impedance and the capacitance of graphene/epoxy coated aluminium alloys were studied using electrochemical impedance spectroscopy (EIS). DSC measurements were performed to evaluate the possible influence of graphene on the curing process of the coating. The resistance of the coating to separation from aluminium alloy substrate was tested by means of a cross-cut test. The hydrophobic characteristics offered by graphene were investigated using contact angle measurements with water sessile drops. The results of the analysis were compared with a coating made of epoxy resin without filler content.

2. EXPERIMENTAL SECTION

2.1. Materials

The graphene nanoflakes were supplied by Cometox (Milano, Italy), with the commercial name of C500. Raman (Bruker, Ettlingen, Germany), X-ray diffraction (XRD; PANalytical, Cambridge, UK) and X-ray photoelectron spectrometer (XPS; Kratos Co., Manchester, UK) analysis of the as-received graphene, shown on the supplier's website, are shown in Figure 1.

In particular, the graphene nanoflakes were 2 μm in diameter, few nanometer in thickness and with an average surface area of 500 m^2/g. One component epoxy system without corrosion inhibitors and a curing agent provided by Sikkens were used as polymer matrix. The mass ratio between the resin and curing agent was 4:1. The metals used were AA 2024-T3 aluminium alloy panels. Before coating the samples, the surfaces are cleaned with ethanol and acetone.

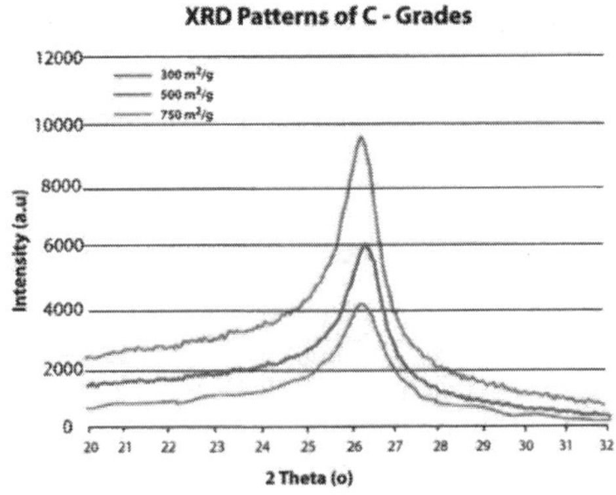

(a)

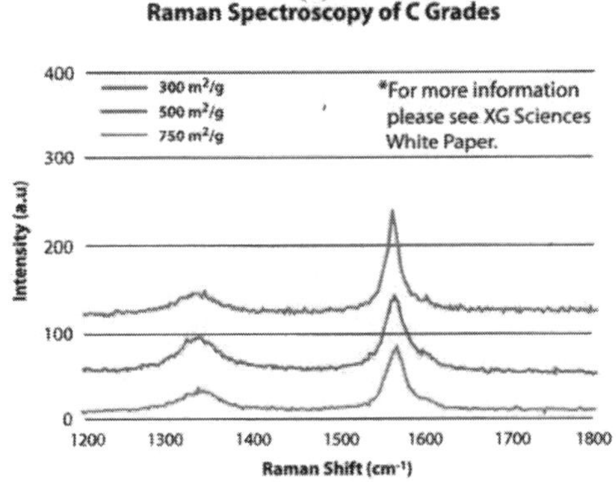

(b)

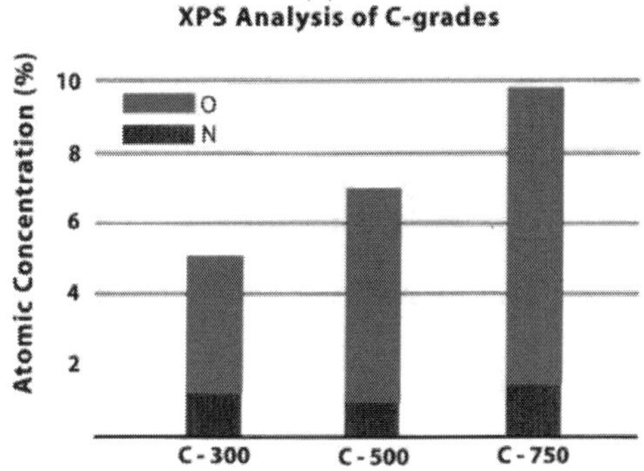

(c)

Figure 1. XRD (**a**), Raman (**b**) and XPS (**c**) of the as-received graphene nanoflakes.

2.2. Graphene/Epoxy Coating Preparation

The epoxy resin and 1 wt% of graphene nanoflakes were mixed. In order to disperse graphene a sonication bath was carried out for 20 min with a frequency of 50 Hz. Next, curing agent was added and mixed using a magnetic stirrer for other 20 min. The hybrid coating was, finally, spread on AA2024 aluminium alloy panel by means of spiral bar applicator and cured at room temperature for 24 h (Figure 2). The thickness of coatings, with or without graphene nanoflakes, measured using an Elcometer thickness instrument (IMCD Italia S.p.A, Milano, Italy), was about 28 ± 1.2 μm.

It can be seen that graphene gives a gray colour to the paint. Certainly this effect can be relevant if the colour of paint becomes an important characteristic.

2.3. Characterizations

2.3.1. Electrochemical Tests

Electrochemical tests were performed in an 3.5 wt% aerated aqueous solution of NaCl, using a conventional 3-electrode system with the graphene-filled coating as the working electrode, a platinum as the counter electrode and a saturated calomel electrode (SCE) as reference electrode. A Solartron 1255 Frequency Response Analyzer (Photoanalytical, Milano, Italy) connected to a Solartron 1286 potentiostat/galvanostat (Photoanalytical, Milano, Italy) was employed. In particular,

(a)

(b)

Figure 2. AA 2024-T3 aluminium alloy coated with epoxy resin (**a**) and graphene/epoxy resin (**b**).

electrochemical impedance spectroscopy (EIS) analysis (Photoanalytical, Milano, Italy) and capacitance tests were performed. The former is a powerful analysis technique to study the corrosion protective properties of coatings over metals. It has been used to characterize the corrosion behaviour of coated aluminium alloys immersed in aggressive solutions [18,19,20].

For EIS measurements, the amplitude of the sinusoidal signal was 10 mV, the frequency ranging from 10^{-2} to 10^{5} Hz. The exposed area was 1 cm^2. Tests were carried out for up to 10 days of immersion. Results were taken by analyzing the Bode plots.

Capacitance tests were performed at open potential circuit (OCP) with an AC amplitude of 40 mV. The excitation frequency was 6.5 kHz. An immersion time of 20 h was chosen.

All the measurements were conducted at room temperature (~25 °C). All trials were repeated at least three times to ensure reproducibility and accuracy of the measurements.

2.3.2. Thermal Analysis

The curing process of both the graphene-filled and unfilled resin was observed by taking 10 mg of sample into aluminum pan. This was placed in sample cell of the instrument, a Mettler Toledo DSC12E (Mettler-Toledo S.p.A., Milano, Italy; in a dry nitrogen atmosphere and calibrated with an Indium standard). Three scans (heating, cooling and heating again) were conducted from 30 to 250 °C with a heating rate of 10 °C/min.

2.3.3. Adhesion Test

The influence of graphene on the adhesion of hybrid coating was examined by cross-cut test, using a cross hatch cutter (Sheen Instruments; Milano, Italy) and following ISO 2409 standard.

2.3.4. Contact Angle Test

Finally, the hydrophobicity of graphene/epoxy coating was determined by analysis of the water contact angle, according to ASTM D7490 and using a contact angle goniometer (Data Physics Ltd., Hailsham, UK). Water drops of 0.2 mL were carefully placed on the substrate with a microlitre syringe. The obtained water contact angle was the average value of ten measurements.

3. RESULTS AND DISCUSSION

3.1. Electrochemical Test

For a perfect insulator (*i.e.*, an intact dry coating without defects), the phase angle should be 90° due to its pure capacitive characteristics. When the electrolytic solution gradually permeates into the material, its phase angle at a given frequency decreases, due to the resistive behaviour assumed by the material and caused by entry of electrolyte. The variation of phase angles may reflect the performance of coatings [21]. In the Bode plot the peaks at different frequency range represent different time constants. The time constant appearing at high frequency range can be attributed to the coating/solution interface, while that at low frequency correspond to a corrosion process taking place at the metal/coating interface [18]. A minimum in the phase angle corresponds to a plateau in the impedance modulus [21], which represents the ability of the coating to impede the flow of current between anodic and cathodic areas.

Figure 3a,b shows the impedance modulus and phase angle of samples coated using unfilled water-based epoxy resin. The former in the medium frequency range exhibited a relevant decrease of about two orders of magnitude after one day. Although it seemed to improve up to 5 days, it decreased again after only 7 days and first blisters appeared on the surface, as shown in Figure 4. When the solution permeates into the epoxy resin, corrosion products are formed, so the impedance decreases (curve 1d). Corrosion products seal the pores of the coating and the impedance restarts to increase (curve 2d). After 5 days the corrosion phenomenon cannot stopped, as confirmed by the trend of impedance modulus.

The damage to the coating was confirmed via the phase angle plot. At beginning of the immersion, the diagram showed two time constants observable as two peaks.

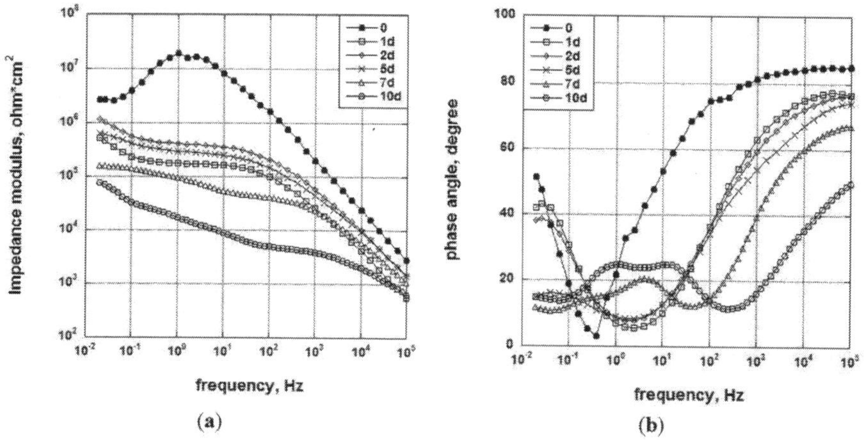

Figure 3. Bode plots for impedance modulus (**a**) and phase angle of sample coated with unfilled water-based epoxy resin (**b**).

The one appearing at high frequency represented the solution/coating interface, while the one at low frequency corresponded to the coating/metal interface, suggesting the presence of pores. In addition, the decrease of peaks height with exposure time indicated that the response became less capacitive; indeed, as immersion time increases, the pores permitted the solution to penetrate into the aluminium substrate, causing localised corrosion. Thus, current flowed through the defects (*i.e.*, localised corrosion) became significant. As immersion time increased, the minimum shifted toward high frequency, suggesting that the property of coating may be affected due to the exposure.

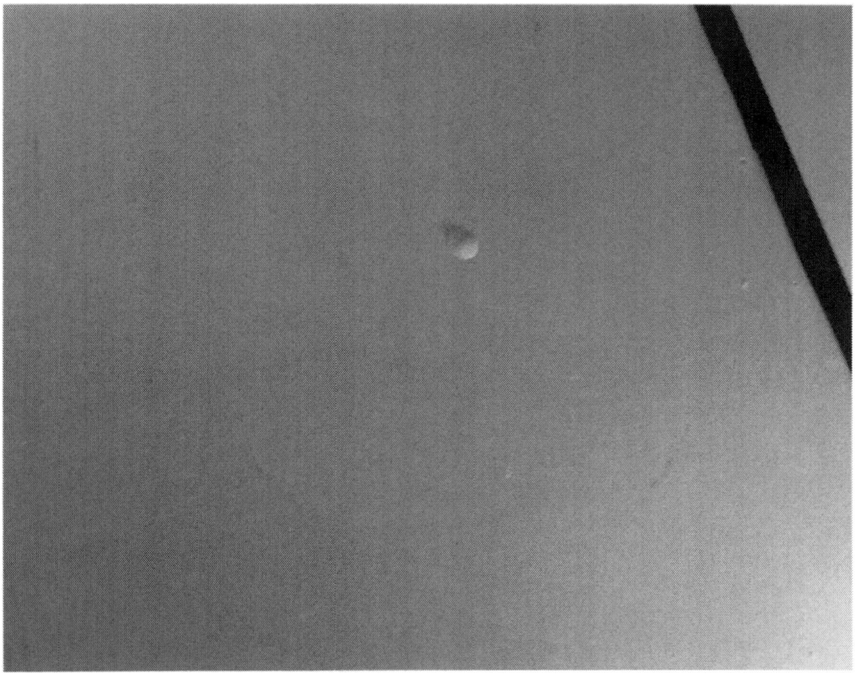

Figure 4. Blisters on sample coated with unfilled water-based epoxy resin after 7 days of immersion in aqueous NaCl solution.

On the other hand, the graphene/epoxy resin no relevant changes were shown throughout 10 days of exposure, as confirmed in Figure 5.

The impedance modulus showed a gradual increase. The angle phase showed one time constant. The height of peak remained unchanging with the immersion time, indicating a stable property of the coating. The impedance spectrum of sample loaded with a small amount of graphene showed an improvement of barrier properties of neat epoxy resin.

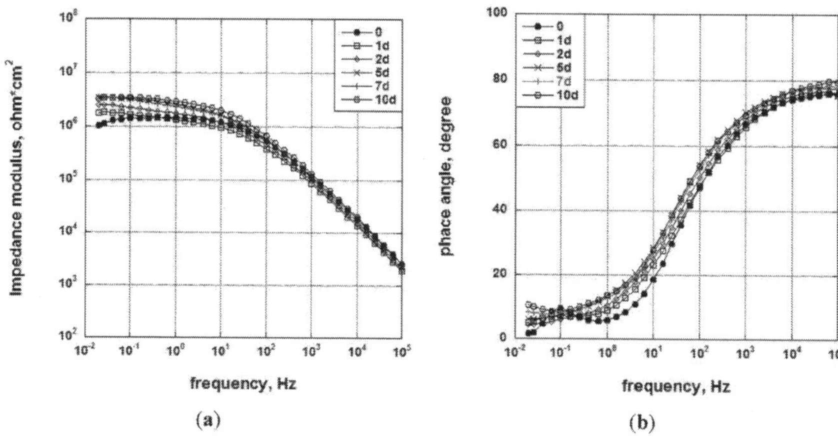

Figure 5. Bode plots for impedance modulus (**a**) and phase angle for sample coated with graphene/epoxy resin (**b**).

The electrochemical characterization also regarded the capacitance test. The results of this analysis were reported in the following figures.

The sample coated with unfilled water-based epoxy resin (Figure 6a), due to very quick water absorption, showed the capacitance response of aluminium substrate. Conversely, the capacitance curve of sample coated with graphene/epoxy resin (Figure 6b) showed a very slow process of water uptake, indicating that there was an improvement of the barrier effect offered by this kind of coating. In fact, no loss of adhesion and no blistering on the aluminium surface were found.

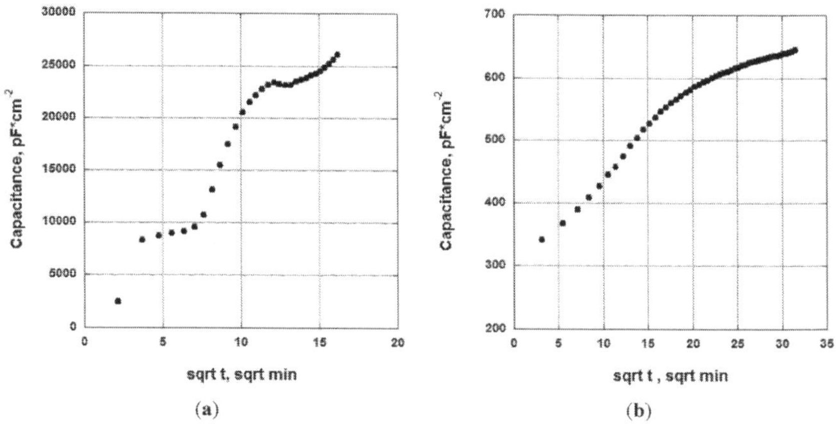

Figure 6. Capacitance curve for sample coated with unfilled water-based epoxy resin (**a**) and graphene/epoxy resin (**b**).

3.2. Thermal Analysis

In order to evaluate the effect of graphene on the curing process of the coating, DSC measurements were carried out (Figure 7).

As can be seen, the water-based epoxy resin not cured (called "epoxy resin liquid") presented a curing peak at approximately 150 °C and another one at about 90 °C, whose nature was not identified. The latter was probably due to some constituents present in the resin.

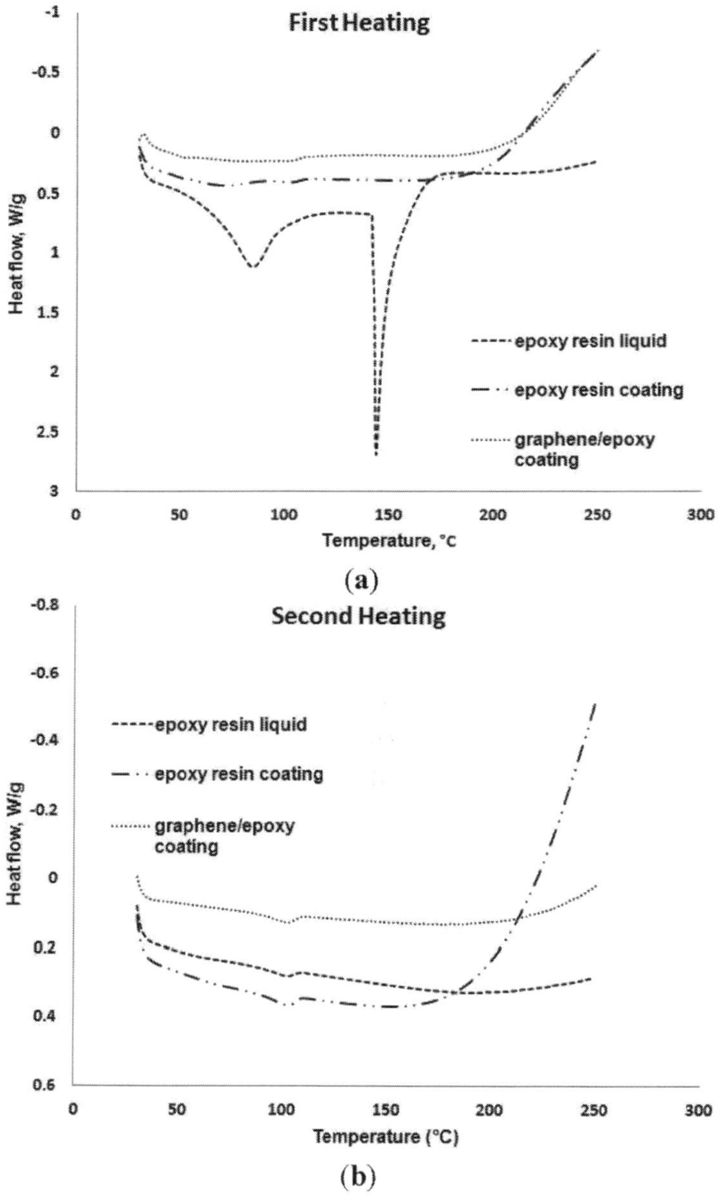

Figure 7. DSC measurements: first heating (**a**) and second heating (**b**).

The water-based epoxy resin that cured (called “epoxy resin coating”) was not shown the curing peak, suggesting that cross-linking reactions were completed during the cure. The same curing profile was displayed by graphene/epoxy coating. This indicated that the presence of graphene did not affect the curing process. Only a small translation was shown due to the different kinds of samples.

3.3. Adhesion Test

The corrosion protection properties of an organic coating most often end up linked to the barrier properties, or to the maintenance of adhesion to the substrate under chemical and electrochemical conditions imposed by the environment. Therefore, in order to investigate the influence of graphene on adhesion of the coating, a cross-cut test was performed, following the ISO 2409 standard.

According to the classification defined in ISO 2409 and ASTM D3002 and D3359, the adhesion of each coating, graphene-filled and unfilled resin (whose pictures were not reported), can be considered sufficient. Thus, the loading of graphene nanoflakes into epoxy resin did not affect the adhesion of hybrid coating.

3.4. Contact Angle Test

In order to analyze the improvement of hydrophobicity of coating, the water contact angle was measured over the surface of graphene-filled and unfilled resin. The obtained values were collected inTable 1.

Table 1. Water contact angle values over graphene-filled and unfilled resin.

Sample	Contact Angle (degree)
Unfilled water epoxy resin	60.4 ± 1
Graphene/epoxy resin	75.3 ± 1

As can be seen in Figure 8, the graphene/epoxy coating showed a higher contact angle, suggesting a minor improvement of barrier properties, as confirmed by capacitance test.

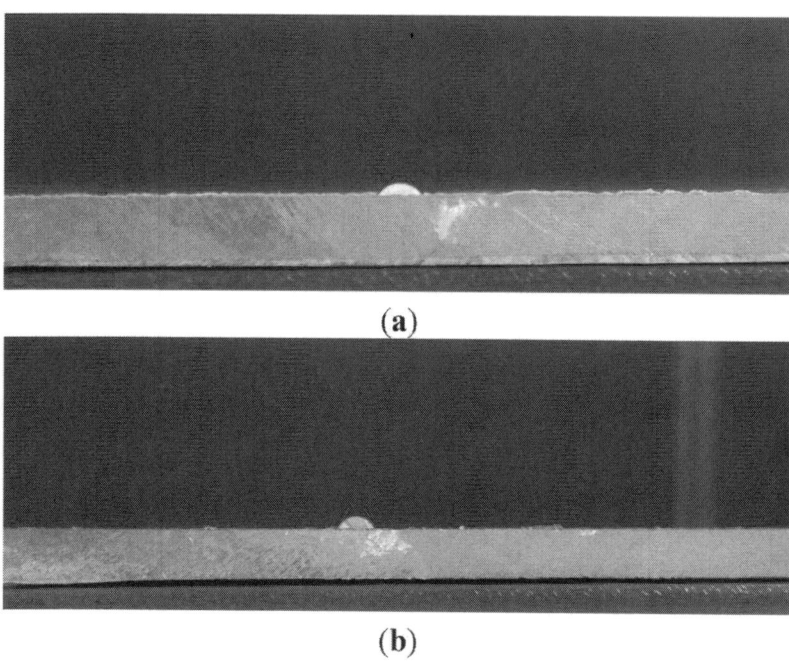

Figure 8. Contact angle of water drops over unfilled water epoxy resin (**a**) and graphene/epoxy resin (**b**).

4. CONCLUSIONS

In this study, graphene/epoxy coating system are prepared by adding a very small amount of graphene nanoflakes to epoxy resin.

- The EIS test demonstrate that the addition of graphene nanoflakes significantly changes the barrier properties of epoxy resin.
- The results of capacitance analysis and water contact angle measurements show that graphene improves the barrier properties of neat epoxy resin.
- The DSC and adhesion tests demonstrate no negligible influence of graphene on the resin curing process or on adhesion properties of the coating, respectively.
- Graphene, added to a non-anticorrosive coating system, has an impressive effect on the corrosion performance of material.

AUTHOR CONTRIBUTIONS

Annalisa Acquesta made all analysis; Tullio Monetta analyzed thermal, adhesion and contact angle tests and Francesco Bellucci analyzed Electrochemical results.

CONFLICTS OF INTEREST

The authors declare no conflict of interest.

REFERENCES

1. Du, J.; Cheng, H.M. The fabrication, properties, and uses of graphene/polymer composites.*Macromol. Chem. Phys.* **2012**, *213*, 1060–1077.

2. Kim, H.; Abdala, A.A.; Macosko, C.W. Graphene/polymer nanocomposite. *Macromolecules* **2010**,*43*, 6515–6530.

3. Kuilla, T.; Bhadra, S.; Yao, D.; Kim, N.H.; Bose, S.; Lee, J.H. Recent advances in graphene based polymer composites. *Prog. Polym. Sci.* **2010**, *35*, 1350–1375.

4. Potts, J.R.; Dreyer, D.R.; Bielawski, C.W.; Ruoff, R.S. Graphene-based polymer nanocomposites.*Polymer* **2011**, *52*, 5–25.

5. Jang, B.Z.; Zhamu, A. Processing of nanographene platelets (NGPs) and NGP nanocomposites: A review. *J. Mater. Sci.* **2008**, *43*, 5092–5101.

6. Yasmin, A.; Luo, J.J.; Daniel, I.M. Processing of expanded graphite reinforced polymer nanocomposites. *Comp. Sci. Technol.* **2006**, *66*, 1182–1189.

7. Zhang, W.; Yi, M.; Shen, Z.; Zhao, X.; Zhang, X.; Ma, S. Graphene-reinforced epoxy resin with enhanced atomic oxygen erosion resistance. *J. Mater. Sci.* **2013**, *48*, 2416–2423.

8. Naebe, M.; Wang, J.; Amini, A.; Khayyam, H.; Hameed, N.; Li, L.H.; Chen, Y.; Fox, B. Mechanical property and structure of covalent functionalised graphene/epoxy nanocomposites. *Sci. Rep.* **2014**,*4*, 4375.

9. Carotenuto, G.; Romeo, V.; Cannavaro, I.; Roncato, D.; Martoraa, B.; Gosso, M. Graphene-polymer composites. *Mater. Sci. Eng.* **2012**, *40*, 012018.

10. Chieng, B.W.; Ibrahim, N.A.; Yunus, W.M.Z.W.; Hussein, M.Z.; Silveraiah, V.S.G. Graphene nanoplatelets as novel reinforcement filler in poly(lactid acid)/epoxidized palm oil green nanocomposites: Mechanical properties. *Int. J. Mol. Sci.* **2012**, *13*, 10920–10934.

11. Mukhopadhyay, P.; Gupta, R.K. Trends and frontiers in graphene-based polymer nanocomposites.*Plast. Eng.* **2011**, *67*, 32–42.

12. Kim, J.Y.; Kim, T.Y.; Suk, J.W.; Chou, H.; Jang, J.H.; Lee, J.H.; Kholmanov, I.N.; Akinwande, D.; Ruoff, R.S. Enhanced dielectric performance in polymer composite films with carbon nanotube-reduced graphene oxide hybrid filler. *Small* **2014**, *10*, 3405–3411.

13. Li, Y.; Yang, Z.; Qiu, H.; Dai, Y.; Zheng, Q.; Li, J.; Yang, J. Self-aligned graphene as anticorrosive barrier in waterborne polyurethane composite coatings. *J. Mater. Chem. A* **2014**, *2*, 14139–14145.

14. Chang, K.C.; Hsu, M.H.; Lu, H.I.; Lai, M.C.; Liu, P.J.; Hsu, C.H.; Ji, W.F.; Chuang, T.L.; Wei, Y.; Yeh, J.M.; *et al.* Room-temperature cured hydrophobic epoxy/graphene composites as corrosion inhibitor for cold-rolled steel. *Carbon* **2014**, *66*, 144–153.

15. Chang, K.C.; Hsu, C.H.; Lu, H.I.; Ji, W.F.; Chang, C.H.; Li, W.Y.; Chuang, T.L.; Yeh, J.M.; Liu, W.R.; Tsai, M.H. Advanced anticorrosive coatings prepared from electroactive polyimide/graphene nanocomposites with

synergistic effects of redox catalytic capability and gas barrier properties.*Express Polym. Lett.* **2014**, *8*, 243–255.

16. Singh, B.P.; Nayak, S.; Nanda, K.K.; Jena, B.K.; Bhattachariee, S.; Besra, L. The production of a corrosion resistant graphene reinforced composite coating on copper by electrophoretic deposition. *Carbon* **2013**, *61*, 47–56.

17. Bellucci, F.; Nicodemo, L.; Nicolais, L. Transport properties of polymeric materials as determined by electrical capacitance method. *Polym. News* **1992**, *17*, 272–277.

18. Bonnel, K.; le Pen, C.; Pébère, N. E.I.S. characterization of protective coatings on aluminium alloys. *Electrochim. Acta* **1999**, *44*, 4259–4267.

19. Bonora, P.L.; Defloriana, F.; Fedrizzi, L. Electrochemical impedance spectroscopy as tool for investigating underpaint corrosion. *Electrochim. Acta* **1996**, *41*, 1073–1082.

20. Zhang, J.; Hu, J.; Zhang, J.; Cao, C. Studies of water transport behavior and impedance models of epoxy-coated metals in NaCl solution by EIS. *Prog. Org. Coat.* **2004**, *51*, 145–151.

21. Juttner, K. Electrochemical impedance spectroscopy (EIS) of corrosion process on inhomogeneous surfaces. *Electrochim. Acta* **1990**, *35*, 1501–1508.

6
CHAPTER

Modelling of the Wireless Propagation Characteristics inside Aircraft

Carl James Debono[1], Reuben A. Farrugia and Keith Chetcuti

[1] *University of Malta, Malta*

1. INTRODUCTION

Advances in wireless communications technology and more sophisticated portable devices have led to a drastic increase in wireless services and applications. This advancement was made possible through hardware improvements which allow more functions to be implemented in smaller sized devices. The demand for more wireless services has pushed the industry and the research community to increase the communication data rates, connectivity, and availability. The increase in the user base has also brought a decrease in the cost of the services.

Although wireless access is becoming available in all public places around the world, this is still not the case for the air passengers who are cut off from wireless services during the duration of the flight. Deployment of wireless technology inside aircraft is still a hot issue due to uncertainties related to interference. Yet, the European Commission has prepared the legal framework for wireless connectivity inside aircraft in April 2008 (Commission, 2008). Aircraft manufacturers will benefit from this technology by exploiting wireless networks to reduce the cable complexity, hence the weight of the aircraft, and by providing new in-flight services, such as online passenger meal selection, service request and video on demand. On the other hand, provisioning of wireless service during flight would profit passengers, since they gain seamless access to common wireless services, such as phone, Internet, and multimedia communications.

A wireless network can always be designed using a measurement campaign, whereby transmitters are placed in strategic locations inside the aircraft and measurements are taken at all points inside the aircraft. This solution is very expensive and requires repetitive readings to measure the propagation losses at all the frequencies lying within the frequency band of interest. These measurements have to be done on all the aircraft models and for every different furniture configuration. While the latter is not much of concern in commercial aircrafts, since these are quite similar in terms of furniture location, this creates a sensible issue in the business jet market, where each aircraft's configuration is tailor-made for the customer. With such requirements, it is not possible to find the ideal antenna locations using measurement techniques.

This calls for the development of an accurate computer model that can estimate the propagation characteristics inside any aircraft. This model must be capable of estimating the power level that can be received at any point inside the enclosed structure, thus creating a propagation map. Therefore, any changes in aircraft model and/or its configuration can be easily accommodated and the propagation map recomputed. This map can then be used to determine the parameters associated with the quality of the wireless channel, hence making the characterisation of the wireless channel possible.

In this chapter we present the application of ray tracing techniques, which are based on geometric optics, to model an IEEE 802.11a wireless system propagation map. This technique has become viable with the increased computational power and speeds available in today's computer technology. The developed propagation model can be used to maximise the coverage within the aircraft cabin and to minimise the power emitted from each access point, by determining the optimum location of its antenna. The cabin presents an interesting scenario because the space within it is limited and there is a high density clutter, because of the furniture and seats, resulting in merely any line-of-sight transmission between the server and the client locations. This implies that propagation inside aircraft is mainly due to reflection, transmission and scattering. Also the cabin does not have any attenuation caused by walls as present in normal office environments. We will also discuss how multipath, coherence bandwidth and time dispersion parameters can be obtained, from the computer model, and used to model the channel. These parameters indicate whether inter-symbol and inter-carrier interference will be experienced by the channels.

2. IEEE 802.11A MODEL

2.1. OVERVIEW OF THE IEEE 802.11 STANDARDS

The IEEE 802.11 standards deployed today are a result of technological advancements, both in hardware and software. The first IEEE 802.11 standard was deployed in 1997 with a maximum throughput of 2Mbps (**Crow, B.P. et al, 1997**). This has gone a long way and current amendments are looking at high speeds of more than 100Mbps (**Paul, T.K. & Ogunfunmi, T. 2008**). The IEEE 802.11 standard has a mandatory throughput of 1Mbps which can be extended to 2Mbps. The standard permits three physical layer implementations, namely, frequency hopping spread spectrum (FHSS), direct sequence spread spectrum (DSSS), and Infra red (IR). The FHSS technique uses Gaussian frequency shift keying (GFSK) for modulation, while DSSS and IR utilise differential binary shift keying (DBPSK) / differential quadrature phase shift keying (DQPSK) and pulse position modulation (PPM) respectively (**Crow, B.P. et al, 1997**).

IEEE 802.11a presents data rates of up to 54Mbps and uses the 5GHz unlicensed national information infrastructure (U-NII) frequency band. The physical layer uses orthogonal frequency division multiplexing (OFDM) multicarrier transmission. The modulation schemes used are binary phase shift keying (BPSK), quadrature phase shift keying (QPSK), and quadrature amplitude modulation (QAM) (IEEE standard, 2003).

2.2. ORTHOGONAL FREQUENCY DIVISION MULTIPLEXING

The application of OFDM to wireless communications has gained ground in the last decade. The main features offered by this technology are: (i) OFDM is spectrally efficient (Can De Beek, J.J. et al, 2002) and (ii) it increases robustness against frequency selective fading (Yomo, H. et al, 2005). The latter is because in a multi-carrier system, it is highly improbable that a fade will affect all the subcarriers and the system can still function.

The orthogonal subcarriers overlap one another making the system bandwidth efficient when compared to non-overlapping multi-carrier systems. This characteristic makes sure that the subcarriers do not interfere with each other. The input stream is divided into N data streams, corresponding to the number of subcarriers, and each subcarrier modulates one of these low rate data streams. A multiplex of these modulated subcarriers results in an OFDM symbol.

One of the problems with OFDM is that orthogonality can be lost if the signal is transmitted over a dispersive channel causing inter-carrier interference (ICI). This problem can be solved by introducing a cyclic prefix (Peled, A. & Ruiz, A., 1980). This solution also prevents inter-symbol interference (ISI) between consecutive OFDM symbols. This means that simple one-tap equalizers, implemented at the receiver, are enough to offer protection against ICI and ISI, making the use of OFDM possible in wireless environments.

2.3. IEEE 802.11A PHY LAYER

The physical (PHY) layer interfaces the medium access control (MAC) layer to the wireless medium. The table below (IEEE standard, 2003) gives a summary of the permissible data rates and their characteristics. A block diagram illustrating how a symbol is generated in the PHY layer is presented in Figure 1.

FIGURE 1. Block diagram of IEEE 802.11a symbol generation.

TABLE 1. IEEE 802.11a physical layer parameters.

Mode	Data rate (Mbps)	Modulation	Coding rate (R)	Coded bits per subcarrier	Coded bits per OFDM symbol	Data bits per OFDM symbol
1	6	BPSK	1/2	1	48	24
2	9	BPSK	3/4	1	48	36
3	12	QPSK	1/2	2	96	48
4	18	QPSK	3/4	2	96	72
5	24	16QAM	1/2	4	192	96
6	36	16QAM	3/4	4	192	144
7	48	64QAM	2/3	6	288	192
8	54	64QAM	3/4	6	288	216

The data coming from the higher layers is first scrambled and protected by the convolutional encoder. Puncturing is applied to these symbols, when necessary, to obtain the output symbol. Symbol interleaving is applied to protect the sequence from burst errors, by introducing sparsity between consecutive data elements. The new data sequence is mapped using the operating modulating scheme and a serial-to-parallel converter divides the data stream into 48 different streams, corresponding to the number of carriers. Pilot signals are then inserted to aid in the demodulation and the channel estimation process at the receiver side. The inverse Fourier transform (IFFT) converts the signal back into the time domain. A cyclic prefix is added to each symbol before re-converting the streams to a single data stream. This cyclic prefix helps in preventing ISI between adjacent symbols and ICI between adjacent carriers. The length of this prefix is a design parameter which depends on the application, since it must be larger than the delay spread of the channel to guarantee that no ISI or ICI distortion effects are introduced in the system.

The OFDM PHY layer divides the information signal across 52 subcarrier frequencies. From these, 48 are used for data transmission and 4 are used by the pilot signals. Another 12 subcarriers are used as cyclic prefix. The available bandwidth is 20MHz and each subcarrier is separated by 312.5kHz. With this configuration, each symbol has a duration of 4s, where the IFFT information has a period of 3.25s while the cyclic prefix uses the remaining 0.75s.

3. PROPAGATION MODEL

The geometry inside an aircraft contains high density clutter, very different from typical office and home scenarios. This means that severe wireless propagation conditions prevail. The advances in computer computation power has made ray tracing modelling possible. Ray tracing techniques offer an adequate representation of the wireless propagation characteristics inside closed, high clutter environments (Diaz, N.R. & Achilli, C., 2003).

The method uses the theory of geometric optics (GO) and is a brute-force solution. Rays are launched from a transmitter at a fixed power level and an estimate of the power levels along the path is found. These rays will undergo reflection, refraction and diffraction as they impinge on the structure of the cabin and the furniture found along the path. The result is a three-dimensional map of the power levels inside the aircraft. This map therefore represents an estimate of the field strength that would be received by mobile terminals at any position inside the aircraft.

3.1. RAY TRACING TECHNIQUE

The two most popular ray tracing techniques are: (i) the method of images, and (ii) the method of shoot and bounce. The first solution gives the exact radiation pattern when the environment is smooth, infinite, or made up of semi-infinite perfect electrically conducting surfaces arranged in a limited set of canonical geometries (Diaz, N.R. & Achilli, C., 2003). The second method is more suitable for complex environments and is therefore used for the propagation modelling inside the aircraft.

In the cabin environment, the signal wavelength, i.e. 5.7cms, is much smaller than the dimensions of the clutter inside the structure and therefore the approximation of the GO hypothesis holds. This implies that ray tracing can be safely used to model the system. A ray is associated with a local plane wave which can be represented by (Diaz, N.R. & Achilli, C., 2003):

$$\nabla^2 \psi + k^2 n^2 \psi = 0 \tag{1}$$

where ψ is the waveform function which governs the scalar wave propagation, k is the wave number, and n is the reflection index of the medium. Solving for ψ gives (Diaz, N.R. & Achilli, C., 2003):

$$\psi = Ae^{-jkS} \tag{2}$$

where A is a function which determines the amplitude of the wave and the function S determines the direction and phase. Using equations (1) and (2), we can conclude that if

$$\frac{\nabla^2}{k^2} n^2 \tag{3}$$

a solution which does not depend on the frequency of operation can be obtained and therefore geometric optics can be used.

As the rays pass through the medium they will experience reflection, refraction and diffraction. This occurs because of the obstacles found in the path between the transmitters and the receivers. If we assume that the obstacles found in the cabin environment are made up of homogeneous material, we simplify the model as each surface can now be described through its dielectric constant, magnetic permittivity and conductivity. This assumption implies that diffraction is not taken into consideration.

If two media possessing different conductivity and permittivity characteristics are assumed to be separated by an infinite plane, then we can obtain equations that relate the reflected electromagnetic wave to the incident wave and the properties of the media. Polarization effects must also be considered and this is done by splitting the electric field into two components; a component which lies parallel to the incident surface and another one which lies perpendicular to it. The refracted and reflected rays can be estimated by multiplying each component to the corresponding Fresnel's coefficient, given by (James, G.L., 1986):

$$R_{\perp} = \frac{\cos\theta - \sqrt{\hat{\varepsilon}_r - \sin^2\theta}}{\cos\theta + \sqrt{\hat{\varepsilon}_r - \sin^2\theta}} \tag{4}$$

$$R_{||} = \frac{\hat{\varepsilon}_r \cos\theta - \sqrt{\hat{\varepsilon}_r - \sin^2\theta}}{\hat{\varepsilon}_r \cos\theta + \sqrt{\hat{\varepsilon}_r - \sin^2\theta}} \tag{5}$$

$$T_{\perp} = \frac{2\cos\theta}{\hat{\varepsilon}_r \cos\theta + \sqrt{\hat{\varepsilon}_r - \sin^2\theta}} \quad (6)$$

$$T_{\parallel} = \frac{2\sqrt{\hat{\varepsilon}_r}\sin\theta}{\hat{\varepsilon}_r \cos\theta + \sqrt{\hat{\varepsilon}_r - \sin^2\theta}} \quad (7)$$

where is the angle of incidence, and $\hat{\varepsilon}_r = \varepsilon_r - j(\eta/2\pi)\sigma\lambda$ is the relative complex dielectric constant. In this equation, r is the relative dielectric constant, is the conductivity of the materials, is the wavelength in meters, R is the reflection coefficients, and T is the refraction coefficients. These relations are valid in this case as the medium is air, which presents a dielectric constant of 0 (Diaz, N.R. & Achilli, C., 2003).

The Fresnel reflection coefficients will account for reflections coming from smooth surfaces. However, this is not the case for the environment being considered. The Rayleigh criterion (Bothias, L., 1987) can be used as a roughness test. When a surface is rough the incident's ray energy will be diffused in angles other than the main angle of reflection, and therefore there is a reduction in the energy of the main reflected ray (Landron, O. et al, 1993). The critical height, hc, in meters, defined by the Rayleigh criterion is given by:

$$h_c = \frac{\lambda}{8 \cos \theta_i} \tag{8}$$

where is the wavelength of the signal and *i* is the angle of incidence.

A surface is considered to be rough if the protuberances exceed h *c*. In such cases the reflection coefficients (R and R ‖) have to be modified by the scattering loss factor (Hashemi, H., 1993):

$$\rho_s = \exp\left[-8\left(\frac{\pi\sigma_h \cos\theta_i}{\lambda}\right)^2\right] I_o \left[8\left(\frac{\pi\sigma_h \cos\theta_i}{\lambda}\right)^2\right] \tag{9}$$

where *h* is the standard deviation of the surface height. The reflection coefficients thus become:

$$(R_\perp)_{rough} = \rho_s (R_\perp)_{smooth} \tag{10}$$

$$(R_{||})_{rough} = \rho_s (R_{||})_{smooth} \tag{11}$$

Within the IEEE 802.11a frequency band, the critical height varies from 7mm at an angle of 1 to 40.9cm at an angle of 89. Inside the cabin, we will not find large roughness in the materials used. Figure 2 shows the profile of the scattering roughness factor for an IEEE 802.11a system. From this figure, it can be seen that the ray can loose up to 75% of its energy when it hits a surface.

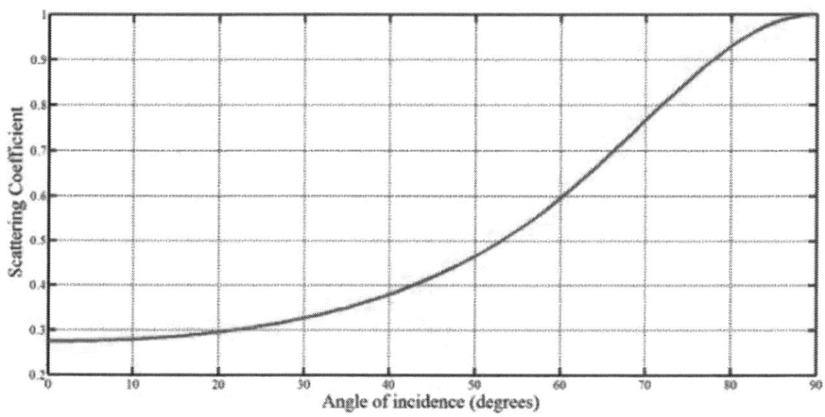

FIGURE 2. Scattering roughness factor against incident angle at 5.25GHz.

3.2. SIGNAL STRENGTH PROPAGATION MODEL

Considering the layout of a typical A340-600 cabin, an aircraft model which includes the furniture is developed. The signal strength at each location inside the aircraft is determined by placing the access points at fixed locations inside the aircraft. Rays are then launched from each transmitting antenna. By vectorially adding all the rays passing through all the points inside the cabin, we obtain an estimate of the signal strength at each position within the aircraft. Therefore, a propagation map which indicates the radio coverage is created.

A ray leaving the transmitter will travel in free space until it impinges on a surface. At this point it is reflected or reflected and refracted as illustrated in figure 3. The rays that result from this interaction are launched again with the new power level from the point of collision. This process will be repeated until the power of the ray falls below a pre-determined threshold.

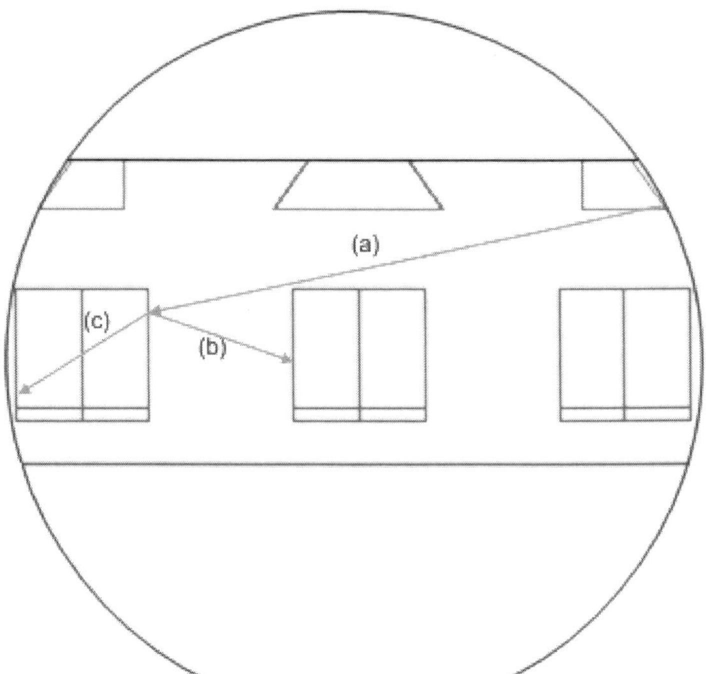

FIGURE 3. The ray tracing technique (a) main ray, (b) reflected ray, and (c) refracted ray.

The power that is received by the path of the kth ray that reaches a single point, is given by (Diaz, N.R. & Achilli, C., 2003):

$$P_k = P_T \left(\frac{\lambda}{4\pi r} \right)^2 G_T G_R \prod_i R_i \prod_j T_j \tag{12}$$

where P$T$is the transmit power in Watts, G$T$and G$R$are the transmitter and receiver gains respectively, is the wavelength in meters, r is the total unfolded path length in meters, R$i$and T$j$are the reflection and refraction coefficients respectively (determined by equations (4) to (7)), and i and j are the indexes that increment over reflection and refraction respectively.

The polarization model can be simplified using techniques found in (Chizhik, D. et al., 1998). The phase, k, of the received field is computed from the fast fading prediction, where $k$is a function of the unfolded path length and the number of reflections. The signal strength at a point in the aircraft can thus be evaluated using:

$$P_R = \left| \sum_k \sqrt{P_k} e^{-j\varphi_k} \right|^2 \tag{13}$$

and

$$\varphi_k = kr + R_{sh} \tag{14}$$

where k is the wave number per meter, r is the length of the path in meters, and R*sh* is the phase shift due to the reflection in radians.

3.3. THE REFLECTION MODEL

Reflection is implemented according to Fermats principle. The direction of the reflected ray is found using:

$$\vec{r} = \vec{v} - 2(\vec{n} \cdot \vec{v})\vec{n} \tag{15}$$

where $\vec{v}$ is the incident ray, $\vec{n}$ is the normal to the plane of incidence, and is the dot product operator. In the cabin environment we will experience a large number of reflections, from every surface. Using Fresnels coefficients, defined above, the rays will experience a phase shift of radians every time there is a reflection. The field power of the reflected signal becomes:

$$P_r = \sqrt{P^2_{r||} + P^2_{r\perp}} \tag{16}$$

where

$$P_{r||} = P_{i||} \cdot \left|R_{||}\right|^2 \tag{17}$$

$$P_{r\perp} = P_{i\perp} \cdot \left|R_{\perp}\right|^2 \tag{18}$$

$$P_{i||} = P_i \cos\theta \tag{19}$$

and

$$P_{i\perp} = P_i \sin\theta \tag{20}$$

The subscripts r and i represent the reflected ray and the incident ray respectively.

The GO principle can also be used to model the propagation of the waves as they hit the curved walls of the aircraft. This can be done because the radius of curvature of this surface is very large compared to the wavelength of the signal. Therefore, the incident ray is reflected at the tangential plane of the surface at the point where the incident ray impinges on the cabin wall.

3.4. THE REFRACTION MODEL

The propagating signal experiences refraction whenever the ray hits an obstacle. The rays which are refracted in a direction of travel which lies outside the aircraft are assumed to be absorbed within the material. This is because we are not interested in the rays which leave the aircraft. The direction of the refracted ray can be calculated using (Diaz, N.R. & Achilli, C., 2003):

$$\vec{t} = n\vec{v} - \left(n(\vec{v},\vec{n}) + \sqrt{1 - n^2\left(1 - (\vec{v},\vec{n})^2\right)} \right) \vec{n} \qquad (21)$$

where n = n *1* /n *2* and n *1* and n *2* are the refraction indexes of the two different media.

The refracted power is calculated using the following:

$$P_t = \sqrt{P_{t||}^2 + P_{t\perp}^2} \qquad (22)$$

where

$$P_{t||} = P_{i||} \cdot |T_{||}|^2 \tag{23}$$

$$P_{t\perp} = P_{i\perp} \cdot |T_{\perp}|^2 \tag{24}$$

where the subscripts i and t represent the incident ray and refracted ray respectively. The model assumes that the obstacles encountered by the rays have constant dielectric properties.

3.5. THE ACCESS POINT MODEL

Each IEEE 802.11a access point is assumed to have an omni-directional antenna. This simplifies things as the access point can be modelled as a point source which radiates the rays uniformly in the three-dimensional space. The Monte Carlo stochastic launching model (Diaz, N.R. & Achilli, C., 2003) is then used to model each transmitter deployed on the aircraft. This will generate rays having random directions within the cabin with equal probability. This ensures that no region within the cabin will contain more rays than another, something which would otherwise skew the results.

The one-dimensional probability density functions are given by (Diaz, N.R. & Achilli, C., 2003):

$$p_\theta(\theta) = \int_0^{2\pi} p_{\theta,\varphi}(\theta,\varphi)d\varphi = \frac{\sin\theta}{2} \tag{25}$$

$$p_\varphi(\theta) = \int_0^{\pi} p_{\theta,\varphi}(\theta,\varphi)d\theta = \frac{1}{2\pi} \tag{26}$$

where and are the spherical coordinates, with 0 2 and 0 . These two randomly distributed variables are generated using:

$$\theta = \arccos\left(1 - 2\xi_1\right) \tag{27}$$

and

$$\varphi = 2\pi\xi_2 \tag{28}$$

where 1 and 2 are random variables which are distributed in [0,1] and in [0,1) respectively.

3.6. MULTIPATH CHARACTERISTICS

We know that the aircraft layout has a very high object density. These objects produce a lot of reflections and refractions as the signals propagate. Therefore, the signal strength arriving at a receiver is highly affected by the large number of multipath signals arriving at that location. These signals will be added vectorially by the receiver hardware. The impulse response can be used to obtain the channel characteristics in these scenarios.

The impulse response of the channel can be modelled using (Hashemi, H., 1993) and (Saleh, A.A.M. & Valenzuela, R.A., 1987). The time invariant impulse response is expressed as a sum of k = 1 .... N multipath components, each having a random amplitude a k, delay k, and phase k:

$$h(t) = \sum_{k=1}^{N} a_k \partial(t - \tau_k) e^{j\theta_k} \tag{29}$$

The three main distributions that are used in communications theory to model multipath effects are the (i) Nakagami, (ii) Rician, and (iii) Rayleigh distributions. At a point inside the aircraft, the signal will experience different fading characteristics as the number of multipath components reaching that point varies. The Rician distribution is more appropriate to scenarios having a dominant line-of-sight signal, which is not the case for the cabin environment. The choice is therefore between the Nakagami and the Rayleigh distribution models. The study in (Can De Beek, J.J. et al, 2002) shows that the multipath distribution of an indoor channel can be better represented by a Nakagami distribution. This distribution is characterised by the cumulative density function:

$$p(x, \mu, \omega) = \frac{2\mu^{\mu}}{\Gamma(\mu)\omega^{u}} x^{2\mu-1} \exp\left(-\frac{\mu}{\omega}x^{2}\right) \quad (30)$$

where is a shape parameter and controls the spread of the distribution.

These distribution parameters have to be extracted from the simulation model. In order to do this, the total number of multipath rays and the maximum and minimum delay times must be recorded for each location inside the cabin. The area inside the cabin is divided into areas, called cells. All the data that is located within the same cell number, which represents the cell distance from the transmitter, is clustered together. The Nakagami model is then fit to this data. Hence, this will give a list of fit parameters that model the multipath propagation inside the cabin.

3.7. TIME DISPERSION PARAMETERS

The IEEE 802.11a channel parameters are characterised by the time dispersion parameters. The main components are composed of: (i) the mean excess delay, and (ii) the root-mean-square (rms) delay spread. These parameters give an estimate of the expected performance that the wireless system will achieve if deployed in the cabin environment. These parameters are then input to the top level IEEE 802.11a system model to obtain the bit error rate (BER) of the channel. The BER results are then used to get an estimate of the quality of service (QoS) and other data transmission characteristics.

The mean excess delay, m, is defined as:

$$\tau_m = \frac{\sum\limits_k \left\{ \left(\tau_k - \tau_1 \right) a_k^2 \right\}}{\sum\limits_k a_k^2} \tag{31}$$

while the rms delay spread, rms, is:

$$\tau_{rms} = \left(\frac{\sum\limits_k \left\{ \left(\tau_k - \tau_m - \tau_1 \right)^2 a_k^2 \right\}}{\sum\limits_k a_k^2} \right)^{\frac{1}{2}} \tag{32}$$

where k is the delay of the kth multipath ray with a normalised amplitude of a k, and 1 is the delay of the line-of-sight signal.

3.8. COHERENCE BANDWIDTH

The coherence bandwidth, B_c, is a measure of the range of frequencies over which two frequency components are likely to have amplitude correlation (Rappaport, T.S., 2002). This bandwidth is related to the rms delay spread. Two signals centered at frequencies that have a separation which is less than or equal to B_c will have similar channel impairments. Otherwise the signals can experience frequency selective fading.

For a frequency correlation function of 0.9 or above, B_c can be approximated by (Lee, W.C.Y., 1989):

$$B_c \approx \frac{1}{50\tau_{rms}} \tag{33}$$

while for a frequency correlation function above 0.5, this approximation becomes:

$$B_c \approx \frac{1}{5\tau_{rms}} \tag{34}$$

Results within a business jet can be found in (Debono, C.J. et al, 2009). For the system to guarantee that the receiver does not experience inter-symbol interference and/or inter-channel interference, the guard interval at any location within the cabin must be less than 800ns (as specified in the IEEE 802.11a standard). Moreover, the system will only introduce flat fading if the coherence bandwidth is greater than the bandwidth of the subcarriers, which is equal to 312.5kHz.

4. SIMULATION RESULTS

4.1. THE CABIN MODEL

A three-dimensional model of the cabin can be developed using any computer aided design (CAD) software. This model can then be imported in the simulation software, which for this work was developed in Matlab®. The propagation characteristics presented here relate to an Airbus A340-600 but further results on a Dassault Aviation business jet can be found in (Debono, C.J. et al, 2009) and (Chetcuti, K. et al, 2009). The structure of the aircraft is modelled through a cylinder which represents the fuselage and a horizontal plane to model the floor. Furniture and the stowage bins were also included as shown in Figure 4.

The seats have a specific thickness and are modelled as two intersecting planes. The dielectric constant, permittivity and conductivity depend on the material used. Typical values of the materials used inside the cabin of the aircraft are given in Table 2.

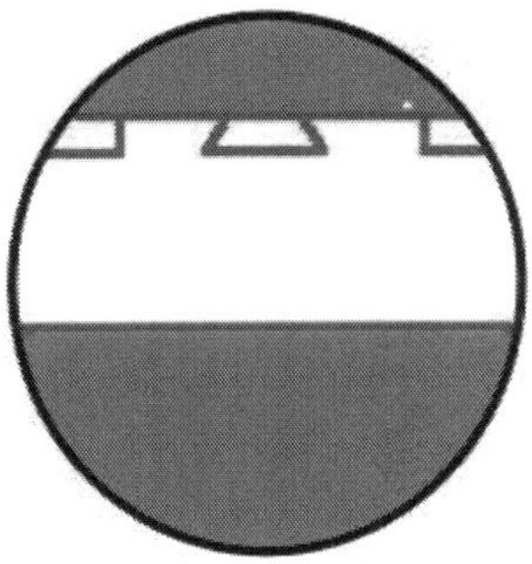

(a)

(b)

FIGURE 4. Cross-section of cabin without seats (a), and with seats (b).

TABLE 2. Electrical characteristics of materials used inside a cabin.

Material	Electric Conductivity	Relative Permittivity
Aluminium	4E7	Inf
Leather	1E-2	3
Wood	1E-2	3

4.2. SIMULATION ENVIRONMENT

The ray tracing algorithm is implemented in Matlab®. A flow chart of the main algorithm is shown inFigure 5. The geometry file used by the developed simulator represents the typical environment of the aircraft under test. The signal strength propagation map is determined by launching 200,000 rays from each transmitter antenna. The equivalent isotropic radiated power from each access point is 30dBm. At any particular cell inside the cabin, the signal strength is determined by summing the power levels of all the rays passing through that point. This implies that the received signal is a distorted version of the transmitted signal.

As discussed in section 3.5, the starting direction of each ray is determined using Monte Carlo techniques, where two random numbers, representing the angles in spherical coordinates, in the range 0 to 2 and 0 to respectively are generated. Each ray is traced one cell size at a time, where at each cell position, the simulator assesses whether the ray is still inside the aircraft. If it is found to lie outside the aircraft, then the trace ends there and the simulator goes back to the antenna to start a new ray trace. If the ray is still inside the aircraft, the propagation loss is calculated. The new power level is compared to the predetermined threshold, which in our case is set to -120dBm, and if it is above this threshold a check is performed to test whether the ray has impinged on a surface. The -120dBm level is well below the minimum detectable signal for IEEE 802.11a, but because of the multipath effects some margin is required to allow for the eventuality that the vectorially summed power level could still exceed the -100dBm limit defined in the standard. The received signal strength at the receiver affects the signal-to-noise ratio (SNR) posing a limit on the maximum useable data rate for error free communication.

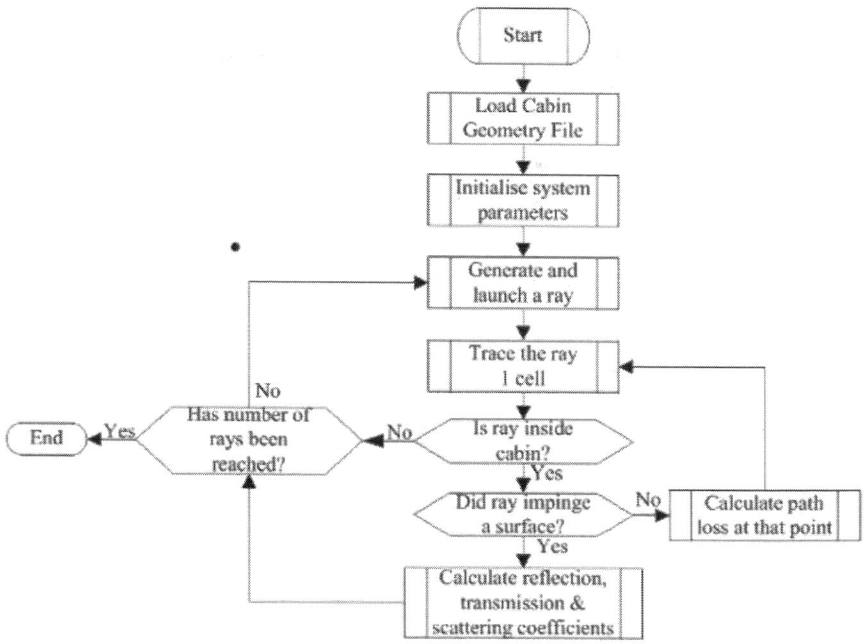

FIGURE 5. Flowchart of the ray tracing method.

4.3. RESULTS

Placing just one access point inside the aircraft limits the number of users that can access the network. This occurs because of the limited capacity that would be offered and the radio propagation coverage that can be obtained with reasonable transmit power levels. The higher the power emanating from the access point, the more interference it is likely to cause to the aircraft's electronics. The simulator developed can be used to determine the optimum number of access points and their position within the aircraft. An analysis for the optimum antenna locations for a Universal Mobile Terrestrial System (UMTS) is found in (Debono, C.J. & Farrugia, R.A., 2008).

The resulting propagation map for the A340-600, using four IEEE 802.11a access points, is shown infigures 6 to 9. Figure 6 presents the view from the antenna plane, Figure 7 shows the top view at the middle of the aircraft, Figure 8 shows the side view, while Figure 9 shows cross-sections looking from the front of the aircraft.

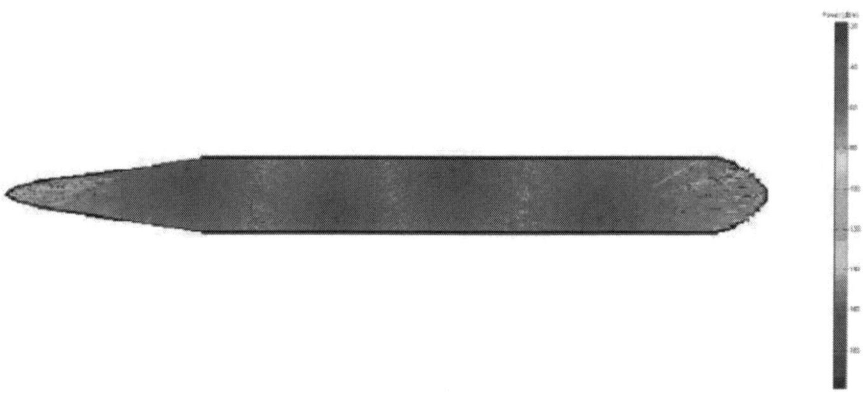

FIGURE 6. Propagation map at the antenna plane. The four access points are shown by the areas of maximum signal concentration.

FIGURE 7. Propagation map at the middle of the aircraft as seen from the top.

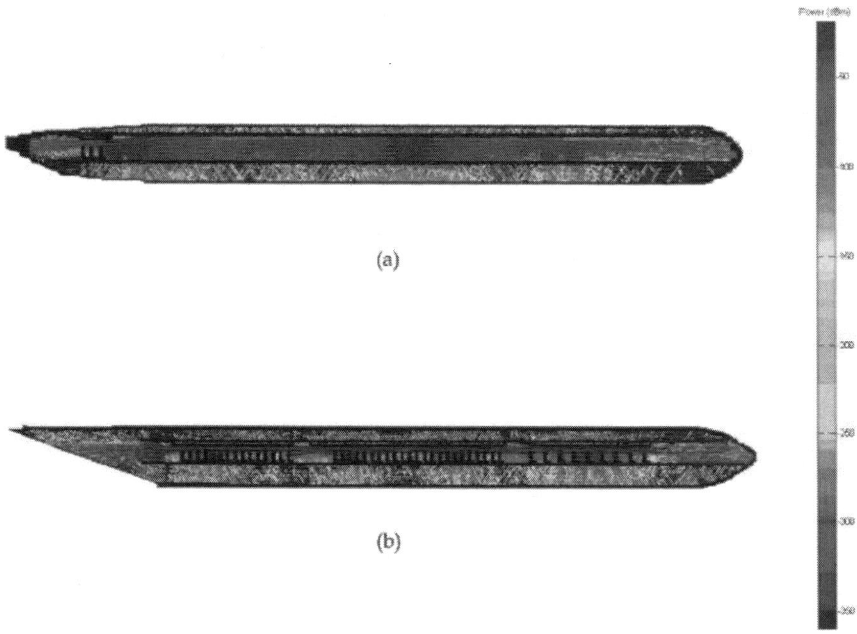

FIGURE 8. The propagation map as seen from the side; (a) aisle, and (b) across a column of seats.

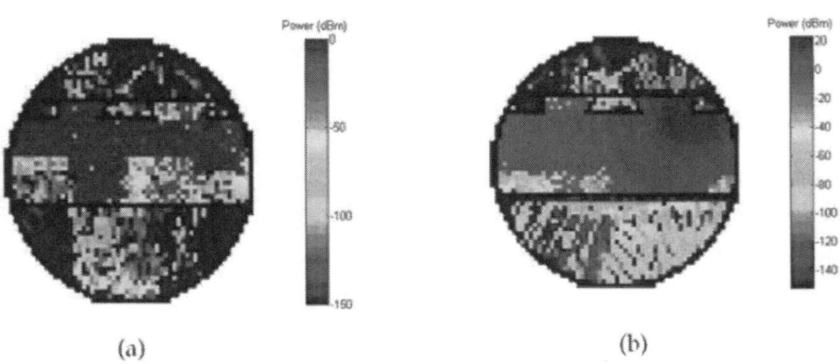

FIGURE 9. The propagation map as seen from the aisle; (a) at the front row, and (b) near one of the access points.

The simulator has also been used to model the propagation inside a business jet. A measurement campaign was done in this case to compare the results obtained and determine the confidence level of the simulations. The results have been presented in (Chetcuti, K. et al, 2009) and show that the model is reasonably accurate, especially within the cabin area.

5. CONCLUSION

This chapter has given the theoretical background necessary to develop a radio propagation map for an IEEE 802.11a system. The method is based on ray tracing techniques which rely on the theory of geometric optics. This solution can be applied to both commercial aircraft, like the A340-600 used in this case, and business jets. The flexibility of the simulator allows easy modifications to simulate different frequency bands and different furniture location and material. This makes the simulator very attractive especially in the business jet environment, where the interior furniture of each aircraft is specifically designed for each customer.

The simulator allows the user to insert the number of access points required and their location. Using an intelligent optimisation technique, such as neural networks, genetic algorithms and support vector machines, one can find the optimum number of access points and their optimum location within the aircraft. This can be done given some constrains imposed by the wiring system of the aircraft.

Moreover, the propagation map gives an idea of the electromagnetic radiation field strength hitting the fuselage of the aircraft. A similar method can be used for each portable device held by each passenger in the aircraft to simulate the uplink.

Therefore, the designer can estimate the electromagnetic interference that is generated by the system. Through optimum design of the system the electromagnetic interference can be kept within acceptable limits and thus ensure that no interference occurs with the aircraft's navigation and control system.

6. ACKNOWLEDGEMENTS

The authors would like to thank Mr. Serge Bruillot from Dassault Aviation for providing us with the model file of their Falcon X7 business jet and for the measurement campaign referenced in the text.

This work forms a small part of the project E-Cab which is financially supported under the European Union 6th Framework Programme (FP6) (E-Cab Website, 2008). The E-Cab consortium is made up of 30 partners from 13 countries across Europe. The authors are solely responsible for the contents of the chapter which does not represent the opinion of the European Commission.

REFERENCES

1. L. Bothias, 1987 Radio Wave Propagation, McGraw Hill Inc., New York, USA.

2. J. J. Can De Beek, P. Odling, S. K. Wilson, P. O. Bojesson, 2002 Orthogonal Frequency-Division Multiplexing, International Union of Radio Science, 2002.

3. K. Chetcuti, C. J. Debono, R. A. Farrugia, S. Bruillot, 2009 Wireless Propagation Modelling Inside a Business Jet, Proceedings of Eurocon 2009, May 2009, 1640 1645 .

4. D. Chizhik, J. Ling, R. A. Valenzuela, 1998 The Effect of Electric Field Polarization on Indoor Propagation, IEEE 1998 International Conference on Universal Personal Communications, October 1998, 459 462 .

5. Commision of the European Communities. 2008 Commission Decision of [...] on harmonised conditions of spectrum use for the operation of mobile communication services on aircraft (MCA services) in the Community- Commision of the European Communities, April 2008.

6. B. P. Crow, I. Widjaja, J. G. Kim, P. T. Sakai, 1997 IEEE 802.11 Wireless Local Area Networks, IEEE Communications Magazine, September 1997, 116 126 .

7. C. J. Debono, R. A. Farrugia, 2008 Optimization of the UMTS Network Radio Coverage On-board an Aircraft, Proceedings of the 2008 IEEE Aerospace Conference, March 2008.

8. C. J. Debono, K. Chetcuti, S. Bruillot, 2009 802.11a Channel Parameters Characterization on board a Business Jet, Proceedings of the 2009 IEEE Aerospace Conference, March 2009.

9. N. R. Diaz, C. Achilli, 2003 Cabin Channel Characterization for Personal Communications via Satellite, Proceedings of the 21st International Communications Satellite Systems Conference and Exhibit, 2003.

10. E-Cab Consortium Website, Online: http://www.e-cab.org

11. H. Hashemi, 1993 The Indoor Radio Propagation Channel, Proceedings IEEE, 81 July 1993.

12. IEEE Std 802.11a-1999(R2003), Part 11: Wireless LAN Medium Access Control (MAC) and Physical Layer (PHY) specifications High-speed Physical Layer in the 5 GHz Band, 2003.

13. G. L. James, 1986 Geometric Theory of Diffraction for Electromagnetic Waves, Peter Peregrinus Ltd., London, UK.

14. O. Landron, M. J. Feuerstein, S. Rappaport, 1993 In Situ Microwave Reflection Coefficient Measurements for Smooth and Rough Exterior Wall Surfaces, Proceedings of the IEEE 43rd Vehicular Technology Conference, May 1993, 77 80 .

15. W. C. Y. Lee, 1989 Mobile Cellular Telecommunications Systems, McGraw Hill Publications, New York, USA.

16. T. K. Paul, T. Ogunfunmi, 2008 Wireless LAN Comes of Age: Understanding the IEEE 802.11n Amendment, IEEE Circuits and Systems Magazine, First Quarter 2008, 8 1 28 54 .

17. A. Peled, A. Ruiz, 1980 Frequency Domain Data Transmission using Reduced Computational Complexity Algorithms, Proceedings of the IEEE International Conference on Acoustics, Speech and Signal Processing, 1980, 964 967 .

18. T. S. Rappaport, 2002 Wireless Communications Principles and Practice, Prentice Hall, New Jersey, USA.

19. A. A. M. Saleh, R. A. Valenzuela, 1987 A Statistical Model for Indoor Multipath Propagation, IEEE Journal on Selected Areas Communications, February 1987, 128 137 .

20. H. Yomo, C. H. Nguyen, P. Kyritsi, T. D. Nguyen, S. S. Chakraborty, R. Prasad, 2005 PHY and MAC Performance Evaluation of IEEE 802.11a WLAN over Fading Channels, Institution of Electronics and Telecommunications Engineers (IETE), 51 1 January 2005, 83 94 .

7

CHAPTER

Concurrent Subspace Optimization for Aircraft System Design

Ke-shi Zhang[1]

[1] Northwestern Polytechnical University, China

1. INTRODUCTION

Concurrent Subspace Optimization (CSSO) is one of the main decomposition approaches in Multidisciplinary Design Optimization (MDO). It supports a collaborative and distributed multidisciplinary design optimization environment among different disciplinary groups. Sobieski first proposed the subspace optimization method (Sobieszczanski-Sobieski, 1988), and Sobieski's blueprint was further developed by Bloebaum and subsequently named the concurrent subspace optimization method (Bolebaum, 1991). Renaud developed a second-order variant of the Global Sensitivity Equation (GSE) method and an alternative potential coordination procedure for the CSSO method (Renaud & Gabriele, 1993a, 1993b, 1994). Sellar proposed to replace GSE with the neutral-network based response surface method (Sellar et al., 1996).

The CSSO method allows a complex couple system to be decomposed into smaller, temporarily decoupled subsystems, each corresponding to different disciplines (subspaces). Each subspace optimization minimizes the system objective function subject to its own constraints as well as constraints contributed from the other subspaces. Each subspace optimization use its own high-fidelity analysis tools as well as given surrogate models or low-fidelity analysis tool provided by the other subspaces for analysis. Subsequently, the subspace optimizations can be performed concurrently. The system-level coordination optimization will be implemented completely based on approximation analysis tools. The subspace optimizations and the coordination optimization will be alternatively performed until results are finally decided by the coordination optimization. Therefore, the CSSO method is particularly suited to applications in a design organization where tasks are distributed among different design groups.

The CSSO method was developed initially for a single objective MDO problem. However, most MDO problems are essentially multi-objective. In recent years more work (Aute & Azarm, 2006; Huang & Bloebaum, 2004; McAllister et al., 2000; McAllister et al., 2004; Orr & Hajela, 2005; Parashar & Bloebaum, 2006; Tappeta & Renaud, 1997; Zhang et al., 2008) has focused on extending existing MDO method to handle such multi-objective MDO problems, by means of integrating a multi-objective optimization method within the MDO framework. This kind of method can be called a multi-objective MDO method.

It is an effective way to integrate multi-objective optimization method within the CSSO framework to develop the multi-objective MDO method. CSSO was extended to solve multi-objective MDO problems, including the Multi-objective Pareto CSSO (MOPCSSO) method, the Multi-objective Range CSSO

(MORCSSO) method, the Multi-objective Target CSSO (MOTCSSO) method, the Multi-objective Genetic Algorithm CSSO (MOGACSSO) method and Adaptive Weighted Sum based CSSO (AWSCSSO). In MOPCSSO the Constraint method is integrated within CSSO framework (Huang & Bloebaum, 2007). In MORCSSO and MOTCSSO the concept of designer preference is introduced (Huang & Bloebaum, 2004). In MOGACSSO the Genetic Algorithm is combined with CSSO and in the hope of improving the computational efficiency (Parashar & Bloebaum, 2006). In AWSCSSO the Adaptive Weighted Sum method is introduced into CSSO (Zhang et al., 2008).

2. GENERAL DESCRIPTION OF MDO PROBLEM FOR AIRCRAFT SYSTEM DESIGN

Aircraft is a complex engineering system in which multiple disciplines (such as aerodynamics, structure, thrust, noise, electronics, cost, et al.) are included. Actually different disciplines are not independent of each other. For example, deformation of a wing structure affects the aerodynamic lift distribution on the wing and in turn a new deformation is caused, which is the well-known aeroelastic problem. The designers of each discipline cannot work without consideration of other disciplines, which makes the aircraft system design become complicated. Especially in the aircraft preliminary design, the specialists of different disciplines have to often work together and discuss with each other to decide many design variables so that higher performance can be achieved. Aircraft system design is a typical multidisciplinary design problem.

In recent years, industry has paid more attention to improving efficiency in the design of complex systems, such as aircraft. MDO has emerged as an engineering discipline that focuses on the development of new design and optimization strategies for the complex systems. MDO researchers strive to reduce the time and cost associated with the coupling interaction among several disciplines. "Decomposition approaches provide many advantages for the solution of complex MDO problems, as they enable a partitioning of a large coupled problem into smaller, more manageable sub-problems. The resulting computational benefits, besides the obvious one associated with the solution of smaller problems, include creating a potential distributed processing environment. The primary benefit, however, pertains to the savings in personal hours, because groups are no longer required to wait around for other groups in the process to complete their design tasks."(Huang & Bloebaum, 2007)

Mathematically for minimization problems the general form for MDO can be represented as follows:

$$
\begin{aligned}
\text{Min}\ & F(X, Y_1, \cdots, Y_N) \\
\text{s.t.}\ & G_i(X, Y_1, \cdots, Y_N) \leq 0 \\
& Y_i = f_i(X, Y_1, \cdots, Y_{i-1}, Y_{i+1}, \cdots, Y_N) \\
& i = 1, 2, \cdots, N \\
& X = \{x_1, x_2, \cdots, x_{NV}\}
\end{aligned}
\tag{1}
$$

Where F is the objective function vector that is composed of one or more objectives, Giis the constraint vector provided by subsystem i, Yiis the coupling vector of subsystem I, and X is the design vector. The objective function and the constraints can be expressed by the function of X and Yi. In this chapter the MDO problem expressed in Eq. (1) will be taken as the example to discuss the CSSO methods.

3. DIFFERENT FRAMEWORKS FOR CONCURRENT SUBSPACE OPTIMIZATION

Global Sensitivity Equation based CSSO and Response Surface based CSSO will be discussed in this chapter as the typical CSSO methods.

3.1. GLOBAL SENSITIVITY EQUATION BASED CSSO (GSECSSO)

GSECSSO is a bi-level optimization method. As an example, the GSECSSO method for a problem with subsystem 1 and 2 (Eq. (1): one objective and two coupled subsystems) is stated in following paragraphs.

The mathematical models of subspace optimizations of GSECSSO can be written as

Sub-optimization 1	Sub-optimization 2
$\text{Min } F(X_1, Y_1, \hat{Y}_2)$	$\text{Min } \hat{F}(X_2)$
$\text{s.t. } C_1(X_1, Y_1, \hat{Y}_2) \le C_1^0(s_1(1-r_1^1)+(1-s_1)t_1^1)$	$\text{s.t. } \hat{C}_1(X_2) \le C_1^0(s_1(1-r_2^1)+(1-s_1)t_2^1)$
$\hat{C}_2(X_1) \le C_2^0(s_2(1-r_1^2)+(1-s_2)t_1^2)$	$C_2(X_2, \hat{Y}_1, Y_2) \le C_2^0(s_2(1-r_2^2)+(1-s_2)t_2^2)$
$Y_1 = f_1(X_1, \hat{Y}_2)$	$Y_2 = f_2(X_2, \hat{Y}_1)$

(2)

Where Xi, Yi and Ci are design variable vector, state variable vector and cumulative constraint of subspace i, respectively. C_i^0 is the value of Ci at the starting point X^0. The value with symbol '^' above is a linearly approximated one. r_k^p, t_k^p and s_p pare responsibility, trade-off and switch coefficients, respectively. In the first iteration, the responsibility coefficients are initialized using the sensitivities of cumulative constraints with respect to design variables (Bolebaum, 1991) and the switch coefficients are set to one. In the following iterations the coordination coefficients are decided by the coordination optimizations.

By means of Kresselmeier-Steinhauser (KS) function (Kreisselmeier & Steinhauser, 1979), the cumulative constraint, Ci, which represents the constraints of subsystem i, is expressed as

$$C_i = g_{i\max} + \frac{1}{\rho}\ln\left(\sum_{k=1}^{m}\exp\left(\rho\left(g_{ik}(X) - g_{i\max}\right)\right)\right) \quad (3)$$

Where $[g_{i1}, g_{i2}, \cdots, g_{im}] = \mathbf{G}_i$ and ρ and ρ is a positive user-prescribed value.

Global Sensitivity Equation (GSE) (Sobieszczanski-Sobieski, 1990) is a method for computing sensitivity derivatives of state (output) variables with respect to independent (input) variables for complex, internally coupled systems, while avoiding the cost and inaccuracy of finite differencing performed on the entire system analysis. By using GSE expressed below, the global sensitivities are calculated, by

$$\begin{bmatrix} I & -\dfrac{\partial Y_1}{\partial Y_2} \\ -\dfrac{\partial Y_2}{\partial Y_1} & I \end{bmatrix} \begin{bmatrix} \dfrac{dY_1}{dX} \\ \dfrac{dY_2}{dX} \end{bmatrix} = \begin{bmatrix} \dfrac{\partial Y_1}{\partial X} \\ \dfrac{\partial Y_2}{\partial X} \end{bmatrix} \tag{4}$$

The responsibility coefficients divide the responsibility of satisfying constraints among all participating subspaces. r_k^p represents the responsibility allocated to the *k*-th subspace for satisfying the *p*-th cumulative constraint. In the first iteration, they are initialized with the sensitivities of cumulative constraints with respect to design variables (Bolebaum, 1991). The trade-off coefficients allow for the violation of a constraint in one subspace optimization in order to gain a large reduction of the objective function. The switch coefficients enable or disable the responsibility coefficients and the trade-off coefficients decided by whether the cumulative constraints are satisfied.

The mathematical model of coordination optimization of GSECSSO can be written as

$$\begin{aligned}
\text{Min} \quad & F(r,t) = F^* + \sum_{p=1}^{2}\sum_{k=1}^{2} \left.\frac{dF}{dr_k^p}\right|_{X=X^*} \Delta r_k^p + \sum_{p=1}^{2}\sum_{k=1}^{2} \left.\frac{dF}{dt_k^p}\right|_{X=X^*} \Delta t_k^p \\
\text{s.t.} \quad & \sum_{k=1}^{2} r_k^p = 1, \quad \sum_{k=1}^{2} t_k^p = 0 \\
& 0 \le r_k^p \le 1, \quad -1 \le t_k^p \le 1, \quad p,k = 1,2
\end{aligned} \tag{5}$$

Where F∗ is the objective value at X∗ (X∗ is the value of design variable vector after subspace optimizations). $\left.\frac{dF}{dr_k^p}\right|_{X=X^*}$ and $\left.\frac{dF}{dt_k^p}\right|_{X=X^*}$ are the optimum sensitivity of F with respect to responsibility and trade-off coefficients.

The flowchart of GSECSSO is shown in Fig. 1. The original problem is decomposed and the mathematical models of subspace optimizations and coordination optimization are established as Eq. (2) and Eq. (5). Based on the initial design X^0, the system analysis is performed and the GSE is used to find the system derivatives. At the first iteration, the design variables are allocated to the appropriate subspaces, as well as the responsibility, trade-off and switch coefficients, *r*, *t*, *s*, are initialized. Subsequently, each subspace optimization minimizes the system objective function subject to its own constraints as well as constraints contributed from the other subspaces. The

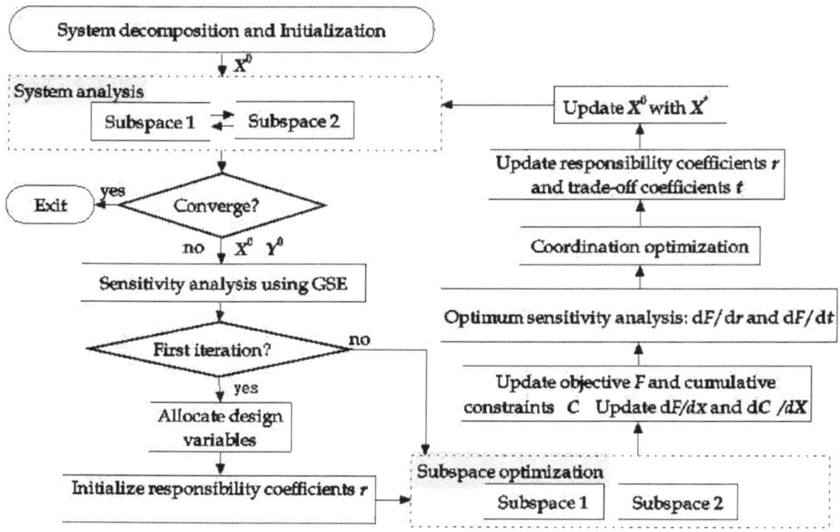

FIGURE 1. Flowchart of GSECSSO method

subspace optimizations are performed concurrently with respect to a disjoint subset of design variables. The updated design vector is the simple combination of local optimal design sub-vectors. After that the global sensitivities and optimum sensitivities are updated. Then the system-level coordination optimization is implemented to optimize coordination parameters.

In GSECSSO, the subspace optimizations are implemented concurrently with respect to a disjoint subset of design variables, which substantially reduces the complexity of the optimization problem within each disciplinary group. The updated design vector is the simple combination of local optimal design sub-vectors. This provides designers with a significant potential benefit in terms of computational effort. However, sometimes the convergence of GSECSSO is oscillatory and premature. It is due to that the trade-off coefficients and the KS parameter ρ with inappropriate value may lead to bad convergence (Huang & Bloebaum, 2004) as main reasons.

3.2. A VARIANT OF GSECSSO

The variant of GSECSSO (Huang & Bloebaum, 2004) is much more efficient than the original GSECSSO. It is adopted in multiple multi-objective CSSO methods (Huang & Bloebaum, 2004;Parashar & Bloebaum, 2006; Zhang et al., 2008).

In the variant of GSECSSO, several modifications are made: (1) the trade-off coefficients are abandoned since all trade-offs will occur directly with minimization of the objective functions within each subspace; (2) The KS parameter, ρ, is set to be increased from a small value; (3) an infeasibility minimization is appended to move the point into the feasible region; (4) the coordination optimization are abandoned and the responsibility coefficients are directly calculated via their initialization strategy.

Taking the MDO problem in Eq. (1) (one objective, two coupled subsystems, i.e. N=2) as an example, the mathematical models of subspace optimizations for the modified CSSO can be written as

$$
\begin{array}{l|l}
\text{Sub-optimization 1} & \text{Sub-optimization 2} \\
\text{Min } F\left(X_1, Y_1, \hat{Y}_2\right) & \text{Min } \hat{F}\left(X_2\right) \\
\text{s.t. } C_1\left(X_1, Y_1, \hat{Y}_2\right) \le C_1^0\left(1 - r_1^1\right) & \text{s.t. } \hat{C}_1\left(X_2\right) \le C_1^0\left(1 - r_2^1\right) \\
\hat{C}_2\left(X_1\right) \le C_2^0\left(1 - r_1^2\right) & C_2\left(X_2, \hat{Y}_1, Y_2\right) \le C_2^0\left(1 - r_2^2\right) \\
Y_1 = f_1\left(X_1, \hat{Y}_2\right) & Y_2 = f_1\left(X_2, \hat{Y}_1\right)
\end{array}
\tag{6}
$$

Where *Xi, Yi* and *Ci* are design variable vector, state variable vector and cumulative constraint of subspace *i*, respectively. C0i is the value of *Ci* at the starting point X^0. The value with symbol '^' above is a linearly approximated one. r_k^p is responsibility coefficients.

the flowchart of modified CSSO method is shown in Fig. 2. In the first stage, based on the initial design variables the system analysis is implemented. If the initial scenario does not satisfy the constraints, a minimization will be required to reduce initial infeasibility of the constraints as much as possible. In the subsequent stage, all the sensitivities are computed by using GSE. Based on the sensitivity information, the impacts of design variables upon each subspace can be analyzed and the design variables will be allocated to the subspace of the greatest impact. In the third stage, the subspace optimizations are performed concurrently. There is no system optimization in this method. The updated design vector is the simple combination of the local optimal design sub-vectors.

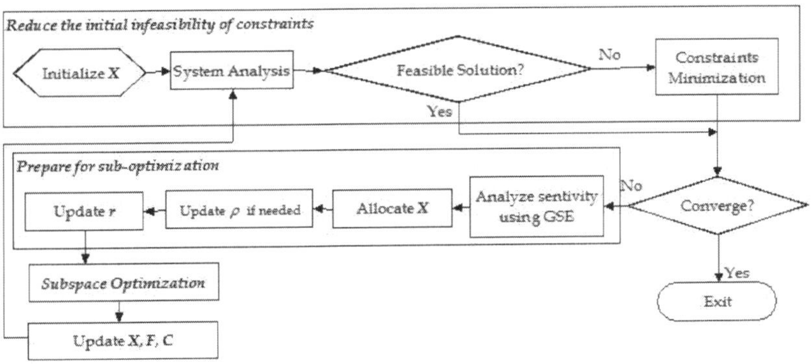

FIGURE 2. Flowchart of the modified CSSO method

The variant of GSECSSO behaves better than GSECSSO on convergence performance. Nevertheless, how to set the start value and increase step of the KS parameter is still hard to say. These two values will affect convergence to some extent and need to be further investigated. The constraint minimization will bring extra computation cost. Furthermore, the linear approximation may sometimes cause oscillatory and premature convergence.

3.3. RESPONSE SURFACE BASED CSSO (RSCSSO)

The RSCSSO method performs optimization in bi-level framework. Taking the MDO problem in Eq. (1) (one objective, two coupled subsystems, i.e. N=2) as an example, the mathematical models of subspace optimizations for RSCSSO can be written as

Sub-optimization 1	Sub-optimization 2
$\text{Min } F\left(X_1, Y_1, \hat{Y}_2\right)$	$\text{Min } F\left(X_2, \hat{Y}_1, Y_2\right)$
$\text{s.t. } G_1\left(X_1, Y_1, \hat{Y}_2\right) \le 0$	$\text{s.t. } G_1\left(X_2, \hat{Y}_1, Y_2\right) \le 0$
$G_2\left(X_1, Y_1, \hat{Y}_2\right) \le 0$	$G_2\left(X_2, \hat{Y}_1, Y_2\right) \le 0$
$Y_1 = f_1\left(X_1, \hat{Y}_2\right)$	$Y_2 = f_2\left(X_2, \hat{Y}_1\right)$

(7)

Where *Xi* and *Yi* are design variable vector and state variable vector of subspace *i*, respectively. In subspace *i*, *Yi* is calculated using the high-fidelity analysis tool while the other state variables with a symbol '^' above are approximated by a quadratic response surface. Actually, other surrogate modeling techniques, such as kriging model or radial basis functions, can be used for approximation as well.

The mathematical model of the coordination optimization of RSCSSO can be written as

$$\begin{aligned} &\text{Min } F\left(X, \hat{Y}_1, \hat{Y}_2\right) \\ &\text{s.t. } G_1\left(X, \hat{Y}_1, \hat{Y}_2\right) \le 0 \\ &\quad\;\; G_2\left(X, \hat{Y}_1, \hat{Y}_2\right) \le 0 \end{aligned} \tag{8}$$

In the system-level optimization, all the state variables are approximately calculated and the design variables of all subspaces are optimized.

The flowchart of RSCSSO is shown in Fig. 3. Firstly, the original problem is decomposed into subspace optimizations in Eq. (7) and coordination optimization in Eq. (8). Then couples of sample points of the design variable vector are generated via Design of Experiment (DOE) method, such as the orthogonal experiment design. The system analysis will be performed based on these data and the results will be stored in the database.

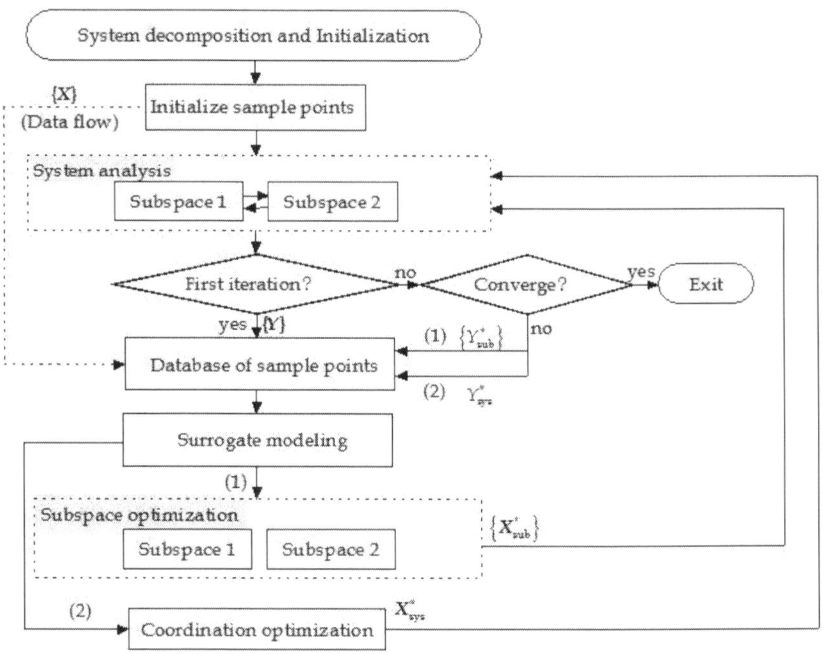

FIGURE 3. Flowchart of RSCSSO method

Subsequently these data will be used to generate the response surface models for the state variables. After that the subspace optimizations are performed simultaneously, the results of which will be analyzed and augmented into the database for updating the response surface models, as flow (1) inFig. 3. Then all of the design variables will be optimized in the coordination optimization and the result will be used to update response surface models as well, as flow (2) in Fig. 3. The subspace optimizations and the coordination optimization are alternately performed until stable convergence is achieved.

In the RSCSSO method, the subspace optimizations are to generate sufficient design information for approximation. After the concurrent subspace optimizations, a global approximation problem is formulated about the current design vector using information stored in the design database. It is the coordination procedure of the global approximation that drives constraint satisfaction and the overall system optimization. A fast and robust convergence appears in the RSCSSO method. However, with an increase in design variables the sample points needed for creating response surface models will be greatly increased. Furthermore, augmenting optimal points is not so reasonable for improving response surface models. As the final result is completely decided by the coordination optimization, the subspace optimizations have little impact on the results and can be abandoned.

3.4. EXAMPLES

3.4.1. EXAMPLE 1: AN ANALYTICAL EXAMPLE WITH COUPLING RELATIONSHIP

The mathematical model for an analytical MDO example considering two coupling disciplines is of the form

$$
\begin{aligned}
\text{Min}\ \ & f = x_2^2 + x_3 + y_1 + e^{-y_2} \\
\text{s.t.}\ \ & g_1 = 8 - y_1 \le 0 \\
& g_2 = y_2 - 10 \le 0 \\
& y_1 = x_1^2 + x_2 + x_3 - 0.2y_2 \\
& y_2 = \sqrt{y_1} + x_1 + x_3 \\
& x_1 \in [-10,10],\ x_2, x_3 \in [0,10]
\end{aligned} \tag{9}
$$

Where y1 and y2 are state variables belonging to subspace 1 and subspace 2, respectively; g1 and g2 are constrains provided by discipline 1 and discipline 2, respectively. The coupling relationship between discipline 1 and discipline 2 is depicted in Fig. 4.

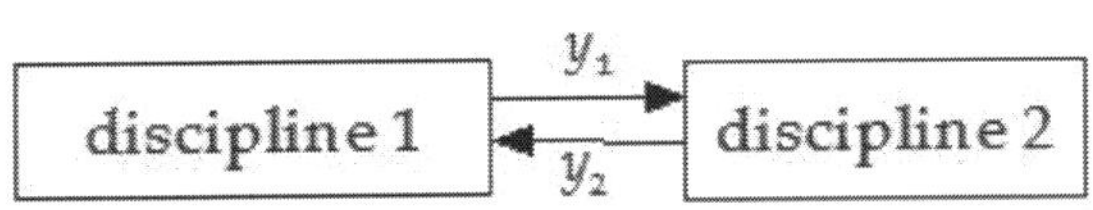

FIGURE 4. Coupling relationship between discipline 1 and discipline 2 for a numerical example

The comparison of the results obtained by different methods is listed in Table 1. Sequential Quadratic Programming (SQP) method is taken as a reference for the convenience of this comparative study. From Table 1, several conclusions can be drawn as follows: (1) the system-level analysis can be remarkably reduced when using different CSSO methods, which shows that the CSSO methods are well suited for MDO problems; (2) the variant of GSECSSO method outperforms the RSCSSO method, in view of the optimization results as well as the required number of system- and disciplinary-level analysis.

TABLE 1. Comparison of different CSSO methods with direct SQP for an analytical optimization problem

variables	SQP	Variant of GSECSSO	RSCSSO
f	8.002860	8.002943	8.031944
x_1	3.028427	3.028442	2.993963
x_2	0.000000	0.000000	0.192359
x_3	0.000000	0.000000	0.000000
Number of system analysis	122	7	38
Number of discipline 1 analysis	0	72	179
Number of discipline 2 analysis	0	64	144

3.4.2. EXAMPLE 2: GEAR REDUCER OPTIMIZATION

This example is taken from the reference (Azarm & Li, 1989). The object is to minimize the overall weight, subject to the constraints for bending and torque stresses. The mathematical model for the optimization problem is as following:

$$
\begin{aligned}
\text{Min} \quad & f = 0.7854x_1x_2^2\left(3.3333x_3^2 + 14.9334x_3 - 43.0934\right) - 1.508x_1\left(x_6^2 + x_7^2\right) \\
& + 7.477\left(x_6^3 + x_7^3\right) + 0.7854(x_4x_6^2 + x_5x_7^2) \\
\text{s.t.} \quad & g_1 = 27x_1^{-1}x_2^{-2}x_3^{-1} - 1 \le 0 \\
& g_2 = 397.5x_1^{-1}x_2^{-2}x_3^{-2} - 1 \le 0 \\
& g_3 = 1.93x_2^{-1}x_3^{-1}x_4^{3}x_6^{-4} - 1 \le 0 \\
& g_4 = 1.93x_2^{-1}x_3^{-1}x_5^{3}x_7^{-4} - 1 \le 0 \\
& g_5 = 10x_6^{-3}\sqrt{\left(745x_2^{-1}x_3^{-1}x_4\right)^2 + 1.69\times 10^7} - 1100 \le 0 \\
& g_6 = 10x_7^{-3}\sqrt{\left(745x_2^{-1}x_3^{-1}x_5\right)^2 + 1.575\times 10^8} - 850 \le 0 \\
& g_7 = x_4^{-1}(1.5x_6 + 1.9) - 1 \le 0 \\
& g_8 = x_5^{-1}(1.1x_7 + 1.9) - 1 \le 0 \\
& g_9 = x_2x_3 - 40 \le 0 \\
& g_{10} = 5 - x_1x_2^{-1} \le 0 \\
& g_{11} = x_1x_2^{-1} - 12 \le 0 \\
& x_1 \in [2.6, 3.6] \quad x_2 \in [0.7, 0.8] \quad x_3 \in [17, 28] \\
& x_4, x_5 \in [7.3, 8.3] \quad x_6 \in [2.9, 3.9] \quad x_7 \in [5.0, 5.5]
\end{aligned}
\tag{10}
$$

The optimization problem is classified into two disciplines: bearing discipline and shaft discipline, without any coupling relationship.

The optimization problem is solved by SQP and different CSSO methods. The results are depicted inTable 2. The convergence histories are shown in Fig. 5. From this comparative study, some conclusions can be drawn as follows: 1) Both GSECSSO and the variant of GSECSSO enable the reduction of system analysis, which in turn improve the efficiency; 2) No reduction of the system analysis is achieved by using RSCSSO. Nevertheless, the optimization procedure as well as the data-flow interface is much simpler, and the optimization can be potentially improved by implementing parallel system analysis for sample points and parallel subspace optimization; 3) The variant of GSECSSO method outperforms the GSECSSO method.

TABLE 2. Comparison of different CSSO methods with direct SQP for gear-box optimization

Variables	SQP	GSECSSO	Variant of GSECSSO	RSCSSO
f	2994.341316	2995.606759	2995.607422	2994.193811
x_1	3.5	3.500000	3.500000	3.500021
x_2	0.7	0.7	0.7	0.700000
x_3	17	17	17	17.000000
x_4	7.3	7.300000	7.3	7.300000
x_5	7.715320	7.733301	7.733330	7.715316
x_6	3.350215	3.352131	3.352156	3.349631
x_7	5.286654	5.287256	5.287246	5.286643
Number of system analysis	40	15	11	51
Number of bearing-discipline analysis	0	326	188	280
Number of shaft-discipline analysis	0	2986	1818	288

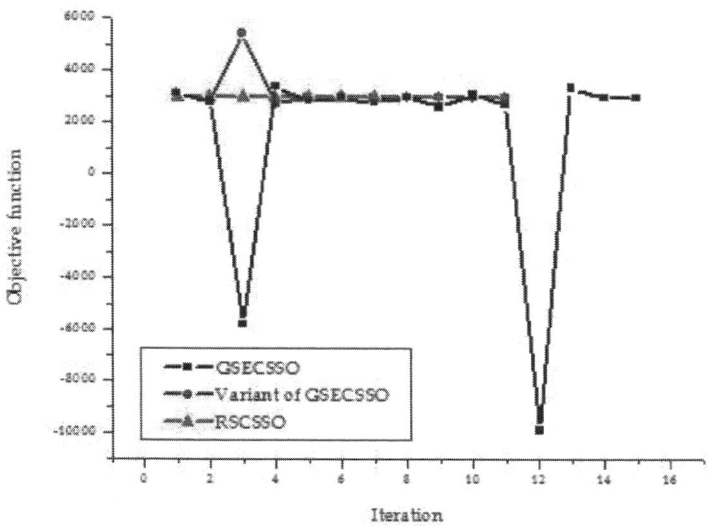

FIGURE 5. Convergence history of CSSO optimization for gear reducer

4. DIFFERENT FRAMEWORKS FOR MULTI-OBJECTIVE CONCURRENT SUBSPACE OPTIMIZATION

The MOPCSSO, MORCSSO and AWSCSSO methods will be discussed in this chapter as the typical multi-objective CSSO methods.

4.1. MULTI-OBJECTIVE PARETO (MOPCSSO)

The constraint method is an effective multi-objective optimization method which optimizes preferred objective with others treated as constraints. MOPCSSO is developed by introducing the constraint method into the variant of GSECSSO.

Taking the MDO problem in Eq. (1) (two objectives, two coupled subsystems, i.e. N=2) as an example, the mathematical models of subspace optimizations of MOPCSSO can be written as

$$
\begin{array}{ll|ll}
\multicolumn{2}{c|}{\text{Sub-optimization 1}} & \multicolumn{2}{c}{\text{Sub-optimization 2}} \\
\text{Min} & F_1\left(X_1, Y_1, \hat{Y}_2\right) & \text{Min} & F_2\left(X_2, \hat{Y}_1, Y_2\right) \\
\text{s.t.} & C_1\left(X_1, Y_1, \hat{Y}_2\right) \le C_1^0\left(1 - r_1^1\right) & \text{s.t.} & \hat{C}_1\left(X_2\right) \le C_1^0\left(1 - r_2^1\right) \\
 & \hat{C}_2\left(X_1\right) \le C_2^0\left(1 - r_1^2\right) & & C_2\left(X_2, \hat{Y}_1, Y_2\right) \le C_2^0\left(1 - r_2^2\right) \\
 & \hat{F}_2 \le F_2^0 & & \hat{F}_1 \le F_1^0 \\
 & Y_1 = f_1\left(X_1, \hat{Y}_2\right) & & Y_2 = f_1\left(X_2, \hat{Y}_1\right)
\end{array}
\tag{11}
$$

The variant of GSECSSO method described in Sub-section 3.2 is adopted in the single objective optimization for each objective function. The flowchart of MOPCSSO is same as that in Fig. 2. During the system optimization, all objective functions are improved simultaneously or individual objective function are improved without worsening the others. The optimization continues until no further improvement can be made so that the Pareto optimum can be obtained finally.

4.2. MULTI-OBJECTIVE RANGE CSSO (MORCSSO)

The goal programming method is one of the most popular multi-objective optimization techniques that consider designer's preference. MORCSSO and Multi-objective Target CSSO (MOTCSSO) are both developed by combining idea of the goal programming method with MOPCSSO.

The $\left[F_2^{\min}, F_2^{\max}\right]$ is supposed to be the preferred range of objective functionF2. The MDO problem in Eq. (1) (two objectives, two coupled subsystems, i.e. N=2) is taken as an example. The framework of MORCSSO is shown in Fig. 6. The variant of GSECSSO method described in Sub-section 3.2 is adopted for sub-optimizations and system-level coordination. The optimization is performed in two stages to obtain a preferred Pareto point.

If the design point does not meet the preference at beginning, the first-stage optimizations are implemented to obtain a design in the preferred range from the starting point.

The mathematical models of subspace optimizations of MORCSSO can be written as

Sub-optimization 1	Sub-optimization 2
$\text{Min } F_1\left(X_1, Y_1, \hat{Y}_2\right)$	$\text{Min } \left\lvert F_2\left(X_2, \hat{Y}_1, Y_2\right) - F_2^{\min}\right\rvert$
$\text{s.t. } C_1\left(X_1, Y_1, \hat{Y}_2\right) \le C_1^0\left(1 - r_1^1\right)$	$\text{s.t. } \hat{C}_1\left(X_2\right) \le C_1^0\left(1 - r_2^1\right)$
$\hat{C}_2\left(X_1\right) \le C_2^0\left(1 - r_1^2\right)$	$C_2\left(X_2, \hat{Y}_1, Y_2\right) \le C_2^0\left(1 - r_2^2\right)$
$-\hat{F}_2 \le -F_2^0$	$Y_2 = f_1\left(X_2, \hat{Y}_1\right)$
$Y_1 = f_1\left(X_1, \hat{Y}_2\right)$	

(12)

If $F_2^0 > F_2^{max}$,

$$\begin{array}{l|l}
\text{Sub-optimization 1} & \text{Sub-optimization 2} \\
\text{Min } F_1\left(X_1, Y_1, \hat{Y}_2\right) & \text{Min } \left|F_2\left(X_2, \hat{Y}_1, Y_2\right) - F_2^{max}\right| \\
\text{s.t. } C_1\left(X_1, Y_1, \hat{Y}_2\right) \le C_1^0\left(1 - r_1^1\right) & \text{s.t. } \hat{C}_1\left(X_2\right) \le C_1^0\left(1 - r_2^1\right) \\
\hat{C}_2\left(X_1\right) \le C_2^0\left(1 - r_1^2\right) & C_2\left(X_2, \hat{Y}_1, Y_2\right) \le C_2^0\left(1 - r_2^2\right) \\
\hat{F}_2 \le F_2^0 & Y_2 = f_1\left(X_2, \hat{Y}_1\right) \\
Y_1 = f_1\left(X_1, \hat{Y}_2\right) &
\end{array} \tag{13}$$

If $F_2^0 < F_2^{min}$,

$$\begin{array}{l|l}
\text{Sub-optimization 1} & \text{Sub-optimization 2} \\
\text{Min } F_1\left(X_1, Y_1, \hat{Y}_2\right) & \text{Min } \left|F_2\left(X_2, \hat{Y}_1, Y_2\right) - F_2^{min}\right| \\
\text{s.t. } C_1\left(X_1, Y_1, \hat{Y}_2\right) \le C_1^0\left(1 - r_1^1\right) & \text{s.t. } \hat{C}_1\left(X_2\right) \le C_1^0\left(1 - r_2^1\right) \\
\hat{C}_2\left(X_1\right) \le C_2^0\left(1 - r_1^2\right) & C_2\left(X_2, \hat{Y}_1, Y_2\right) \le C_2^0\left(1 - r_2^2\right) \\
-\hat{F}_2 \le -F_2^0 & \hat{F}_1 \le F_1^0 \\
Y_1 = f_1\left(X_1, \hat{Y}_2\right) & Y_2 = f_1\left(X_2, \hat{Y}_1\right)
\end{array} \tag{14}$$

In the second stage the design point is optimized closer to the Pareto frontier gradually within the preferred range.

After the optimization in the first stage, the design is in the designer's preferred objective range. Then the mathematical models of subspace optimizations of MORCSSO is same as those of MOPCSSO in Eq. (11).

In the course of optimization in the second stage, the design point maybe flees out of the preferred objective range again. In such a case, the optimization below should be performed.

If $F_2^0 > F_2^{max}$,

Sub-optimization 1	Sub-optimization 2
Min $F_1(X_1, Y_1, \hat{Y}_2)$	Min $\left|F_2(X_2, \hat{Y}_1, Y_2) - F_2^{max}\right|$
s.t. $C_1(X_1, Y_1, \hat{Y}_2) \le C_1^0(1 - r_1^1)$	s.t. $\hat{C}_1(X_2) \le C_1^0(1 - r_2^1)$
$\hat{C}_2(X_1) \le C_2^0(1 - r_1^2)$	$C_2(X_2, \hat{Y}_1, Y_2) \le C_2^0(1 - r_2^2)$
$\hat{F}_2 \le F_2^0$	$\hat{F}_1 \le F_1^0$
$Y_1 = f_1(X_1, \hat{Y}_2)$	$Y_2 = f_1(X_2, \hat{Y}_1)$

(15)

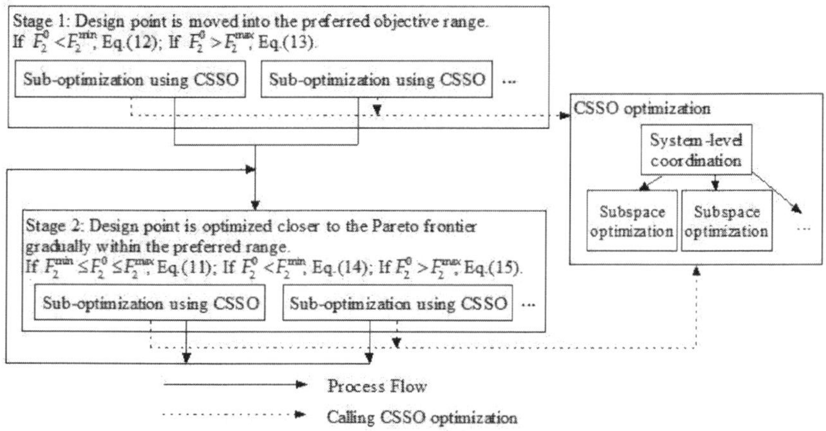

FIGURE 6. Framework of MORCSSO method

For the MORCSSO method, in the case of $F_2^0 > F_2^{max}$ in the second-stage optimizations, the optimizations may converge to the Pareto frontier above the preferred objective range, i.e. the Pareto point that $F_2 > F_2^{max}$ is obtained.

Why does the optimization fail in this case? It can be analyzed from Eq. (15) in the objective space shown in Fig. 7. For Eq. (15), (1) X0 should be closer to the line $F_2 = F_2^{max}$ according to Min | $\left| F_2 - F_2^{max} \right|$, (2) $F_2(X_0)$, (2) $F_2(X_0)$ should be no more than F_2^0 according to $\hat{F}_2 \le F_2^0$, (3) X_0, (Eq. 3) X0 should optimized to decrease F_1^0 according to $\hat{F}_1 \le F_1^0$

And Min F_1, (4) X0X0should be in the feasible region. As shown in Fig. 7, (2) and (3) forces X0X0 to move along a direction of n⃗ 1n→1 in the lower-left shadowed region, in which direction the design point will impossibly move into the preferred objective range. Only along a direction of $\vec{n}_2$ in the lower-right shadowed region the preferred range can be achieved. In such a case X0 may move down straightly along the direction of $\vec{n}$ to arrive at the Pareto front. The semi-infinite region between $F_1 \leq F_{\max 1}$ and the Pareto front is named as Blind Region in this chapter, which means the point falling into this region will not converge to the Pareto front in the preferred region any more. This error will happen in the case of three and more objectives as well, as shown in Fig. 8.

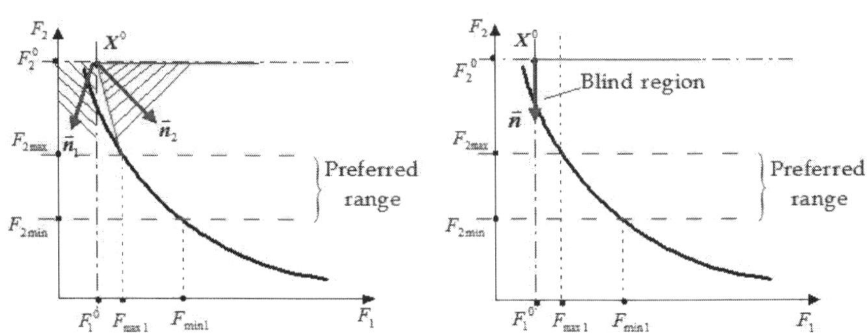

FIGURE 7. The analysis of Eq. (15) in bi-objective space

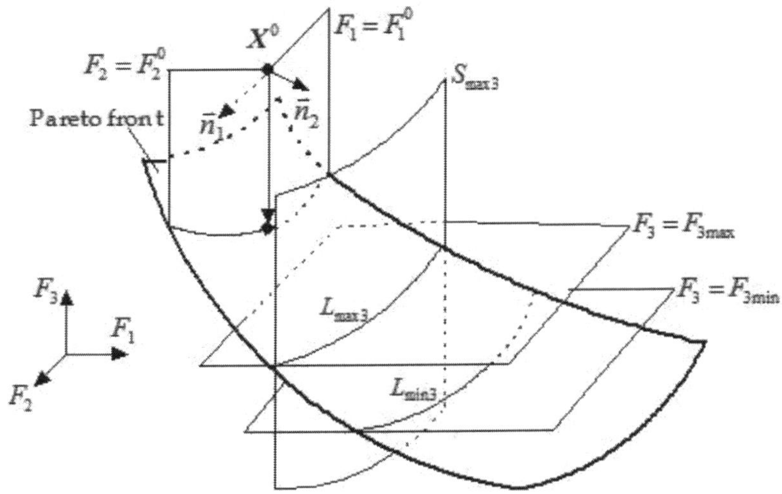

FIGURE 8. The analysis of Eq. (15) in three-objective space

How to solve this problem? From Fig. 7, we only need to move the line $F_1 = F_1^0$ right a little bit, then the two shadowed region will be crossed each other. In the mathematical meaning $F_1 \leq F_1^0$ is relaxed to

$F_1 \leq F_1^0 + \breve{\omega}\left|F_1^0\right|$ in Eq. (15). This strategy is proven to be effective.

4.3. ADAPTIVE WEIGHTED SUM BASED CSSO (AWSCSSO)

The procedure of solving the Pareto front by AWSCSSO is similar to Adaptive Weighted Sum (AWS) method (Kim & de Weck, 2004, 2005). As an example, the AWSCSSO method for a generic bi-objective problem, with subsystem 1 and 2, is stated in the following paragraphs (Eq. (1): two objectives and two coupled subsystems). In the same way AWSCSSO can be also applied for the multi-objective problem with three or more subsystems.

In the first stage a rough profile of the Pareto front is determined.

The variant of CSSO method described in Sub-section 3.2 is adopted in the single objective optimization for each objective function and objective function is normalized as following:

$$\overline{J}_i = \frac{J_i - J^{\text{Nadir}}}{J^{\text{Utopia}} - J^{\text{Nadir}}} \tag{16}$$

When X^{i*} is the optimal solution vector for the single objective optimization of the *i*th objective function *Ji*, the utopia point and pseudo nadir point are defined as

$$J^{\text{Utopia}} = \left[J_1\left(\mathbf{X}^{1^*}\right), J_2\left(\mathbf{X}^{2^*}\right), \cdots, J_{\text{m}}\left(\mathbf{X}^{\text{m}^*}\right) \right] \tag{17}$$

$$J^{\text{Nadir}} = \left[J_1^{\text{Nadir}}, J_2^{\text{Nadir}}, \cdots, J_{\text{m}}^{\text{Nadir}} \right] \tag{18}$$

Where $J_i^{\text{Nadir}} = \max\left[J_i\left(\mathbf{X}^{1^*}\right) \; J_i\left(\mathbf{X}^{2^*}\right) \; \cdots \; J_i\left(\mathbf{X}^{\text{m}^*}\right) \right]$

and m is the number of objective functions.

Then with a large step size of the weighting factor the usual weighted sum method is used in the variant of GSECSSO to approximate the Pareto front quickly. The subspace optimization for AWSCSSO can be expressed as

Sub-optimization 1	Sub-optimization 2
Min $W_1 F_1(X_1, Y_1, \hat{Y}_2) + W_2 \hat{F}_2(X_1)$	Min $W_1 \hat{F}_1(X_2) + W_2 F_2(X_2, \hat{Y}_1, Y_2)$
s.t. $C_1(X_1, Y_1, \hat{Y}_2) \le C_1^0(1 - r_1^1)$	s.t. $\hat{C}_1(X_2) \le C_1^0(1 - r_2^1)$
$\hat{C}_2(X_1) \le C_2^0(1 - r_1^2)$	$C_2(X_2, \hat{Y}_1, Y_2) \le C_2^0(1 - r_2^2)$
$Y_1 = f_1(X_1, \hat{Y}_2)$	$Y_2 = f_2(X_2, \hat{Y}_1, \hat{Y}_3)$

(19)

Where the value with symbol '^' above is a linearly approximated one, C_1 and C_2 are cumulative constraints of G_1 and G_2, respectively, and r_k^p represents responsibility assigned to the k-th subsystem for reducing the violation of Cp. The value with superscript '0' is corresponding to the starting pointX_0. W_1 and W_2 are weighting factors for objective function vector F_1 and F_2, respectively.

By estimating the size of each Pareto patch, the refined regions in the objective space are determined. An example of the mesh refinement in AWSCSSO is shown in Fig. 9. Where hollow points represent the newly refined node P_E (expected solution) while solid points represent initial four nodes that define the patch. As shown in Fig. 9, the quadrilateral patch is taken as an example. If the line segment that connects two neighboring nodes of the patch is too long, it is divided into only two equal line segments. The central point becomes the

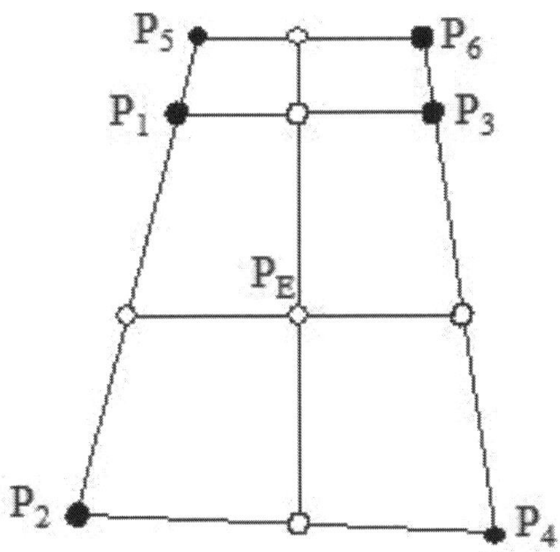

FIGURE 9. Refine patches of AWSCSSO method

new refined node. These refined nodes are connected to form a refined mesh. Then the sub-optimizations in Eq. (19) are performed using different additional constraints for different refined nodes and the new Pareto points are obtained. In next step, according to the prescribed density of Pareto points, the Pareto-front patch that is too large will be refined again in the same way. In subsequent steps, the refinement and sub-optimizations are repeated until the number of Pareto points does not increase anymore.

In the subsequent stage only these regions are specified as feasible domains for sub-optimization problem with additional constraints. Each Pareto front patch is then refined by imposing additional equality constraints that connect the pseudo nadir point (P_N) and the expected Pareto optimal solutions (P_E) on a piecewise planar surface in the objective space (as shown in Fig. 10).

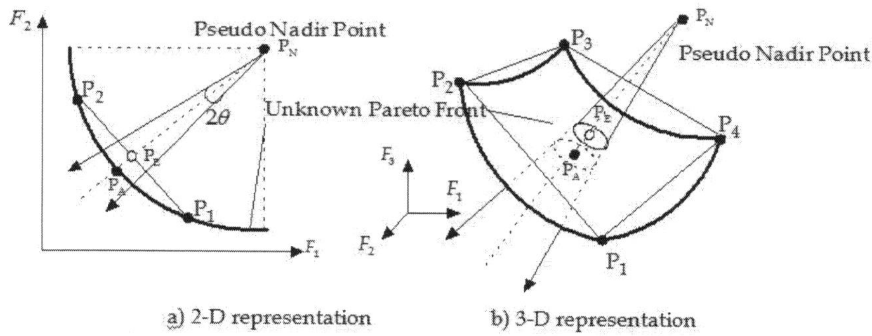

FIGURE 10. AWSCSSO method for multidimensional problems

Sub-optimizations are defined by imposing additional constraint H to Eq. (19) as

Sub-optimization 1	Sub-optimization 2
Min $W_1 F_1(X_1, Y_1, \hat{Y}_2) + W_2 \hat{F}_2(X_1)$	Min $W_1 \hat{F}_1(X_2) + W_2 F_2(X_2, \hat{Y}_1, Y_2)$
s.t. $C_1(X_1, Y_1, \hat{Y}_2) \le C_1^0(1 - r_1^1)$	s.t. $\hat{C}_1(X_2) \le C_1^0(1 - r_2^1)$
$\hat{C}_2(X_1) \le C_2^0(1 - r_1^2)$	$C_2(X_2, \hat{Y}_1, Y_2) \le C_2^0(1 - r_2^2)$
$Y_1 = f_1(X_1, \hat{Y}_2)$	$Y_2 = f_2(X_2, \hat{Y}_1, \hat{Y}_3)$
$H(X_1, Y_1, \hat{Y}_2) \le 0$	$H(X_2, \hat{Y}_1, \hat{Y}_3) \le 0$

(20)

The additional inequality constraint is

$$H = -\frac{\left(\bar{F}_{\mathrm{E}} - \bar{F}_{\mathrm{N}}\right)\cdot\left(\bar{F}(X) - \bar{F}_{\mathrm{N}}\right)}{\left|\bar{F}_{\mathrm{E}} - \bar{F}_{\mathrm{N}}\right|\left|\bar{F}(X) - \bar{F}_{\mathrm{N}}\right|} + \mathrm{L} \le 0 \tag{21}$$

Where L is the adaptive relax factor that is less than 1. $\bar{F}_{\mathrm{E}}$, $\bar{F}_{\mathrm{N}}$ and $\bar{F}(X)$ are the normalized position vector of node P_E, P_N and the current design point X respectively. In AWSCSSO, L is set to be increased with the rise of the distribution density of Pareto points.

Fig. 11 shows the framework of AWSCSSO. In Fig. 11, W^{1i} and W^{2i} are weighting factors in stage 1 and stage 2, respectively; H is the additional constraint. The optimization problem is performed in two stages in the AWSCSSO method. In the first stage the Pareto front is approximated quickly with large step size of weight factors. The optimization problems of this stage are defined in Eq. (19). In the subsequent stage, by calculating the distances between neighboring solutions on the front in objective space, the refined regions are identified and the refined mesh is formulated. Only these regions then become the feasible regions for optimization by imposing additional constraints in the objective space. The optimization problems of this stage are defined in Eq. (20). The different locations of new Pareto points are defined by the different additional constraints. Optimization is performed in each of the regions and the new solution set is acquired. Being a MDO problem, the optimization is performed by the variant of GSECSSO method.

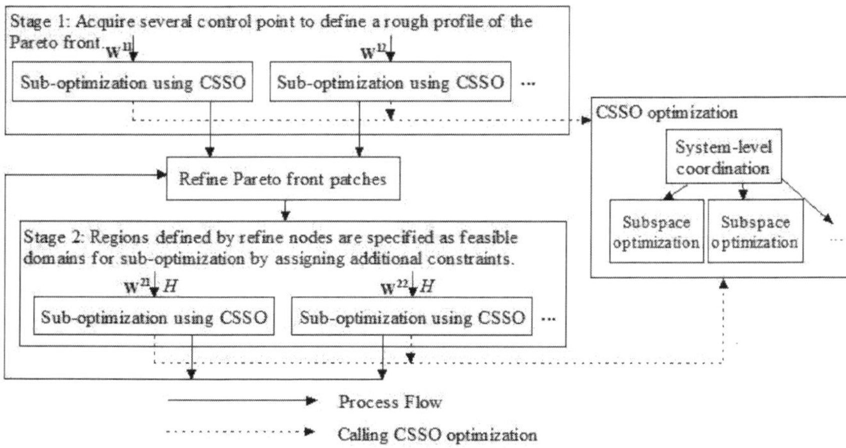

FIGURE 11. Framework of AWSCSSO method

4.4. EXAMPLES

4.4.1. EXAMPLE 1: CONVEX PARETO FRONT

This problem is taken from a test problem (Huang, 2003). This is a test problem available in the NASA Langley Research Center MDO Test Suite. It has two objectives, F_1 and F_2, to be minimized. It consists of ten inequality constraints, four coupled state variables and ten design variables in two coupled subsystems. The mathematical model is not listed here for concision. We refer the readers to the test problem 1 in the corresponding references.

The comparison of the solution obtained by MOPCSSO and AWSCSSO is shown in Fig. 12. It can be concluded that for the problem with convex Pareto front a uniformly-spaced wide-distributed and smooth Pareto front can be obtained by AWSCSS method. When using MOPCSSO I have not captured the whole range.

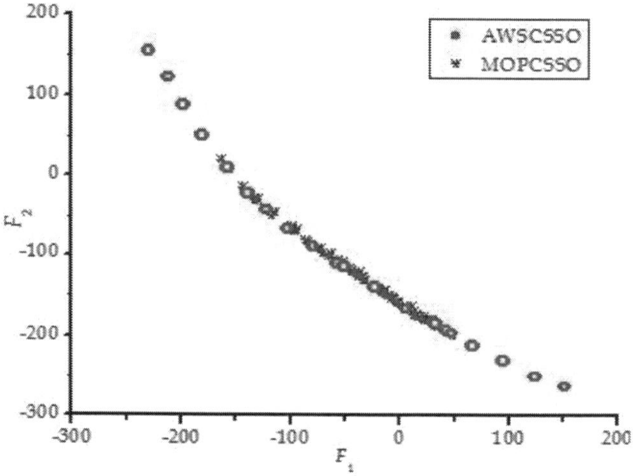

FIGURE 12. Comparison of Pareto front obtained by using AWSCSSO and MOPCSSO

4.4.2. EXAMPLE 2: NON-CONVEX PARETO FRONT

This problem consists of two objective functions, six design variables and six constraints. Two objectives, F_1 and F_2 need to be minimized. The model problem is defined as

$$
\begin{aligned}
&\text{Min} \;\; F_1(x) = -\left(25(x_1-2)^2 + (x_2-2)^2 + (x_3-1)^2 + (x_4-4)^2 + (x_5-1)^2\right) \\
&\text{Min} \;\; F_2(x) = x_1^2 + x_2^2 + x_3^2 + x_4^2 + x_5^2 + x_6^2 \\
&\text{s.t.} \;\;\; c_1(x) = x_1 + x_2 - 2 \geq 0 \\
&\qquad\; c_2(x) = 6 - x_1 - x_2 \geq 0 \\
&\qquad\; c_3(x) = 2 + x_1 - x_2 \geq 0 \\
&\qquad\; c_4(x) = 2 - x_1 + 3x_2 \geq 0 \\
&\qquad\; c_5(x) = 4 - (x_3-3)^2 - x_4 \geq 0 \\
&\qquad\; c_6(x) = (x_5-3)^2 + x_6 - 4 \geq 0 \\
&\qquad\; 0 \leq x_1, x_2, x_6 \leq 10, \quad 1 \leq x_3, x_5 \leq 5, \quad 0 \leq x_4 \leq 6
\end{aligned} \tag{22}
$$

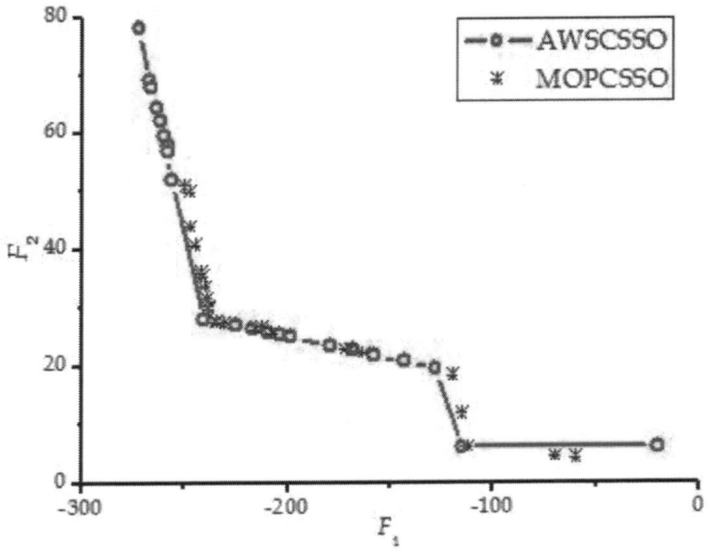

FIGURE 13. Comparison of Pareto fronts obtained by using AWSCSSO and MOPCSSO methods

The comparison of Pareto front obtained by AWSCSSO and MOPCSSO is shown in Fig. 13. It is concluded that, for the problem with non-convex Pareto front, the more uniformly-spaced, more widely-distributed and smoother Pareto front is also obtained by the AWSCSSO method.

4.4.3. EXAMPLE 3: CONCEPTUAL DESIGN OF A SUBSONIC PASSENGER AIRCRAFT

The mathematical model of this problem is defined as

$$\begin{aligned} &\text{Max } U \\ &\text{Max } L/D_C \\ &\text{s.t.} \quad C_{d0L} \le 0.2,\ C_{d0C} \le 0.02 \\ &\qquad R_f \ge 1 \\ &\qquad q_{To} \ge 0.027,\ q_L \ge 0.024 \\ &\qquad D_{To} \le 1981,\ D_L \le 1371 \end{aligned} \tag{23}$$

The objective functions in Eq. (23) are to maximize useful load fraction (U) and lift-to-drag ratio for the cruising condition (L/D_C). The constraints in Eq.(23) are as follows. (1) The drag coefficient for the take-off condition and landing condition (C_{d0L}) is no more than 0.2 and that for the cruising condition (C_{d0C}) is no more than 0.02. (2) The overall fuel balance coefficient (R_f) is no less than 1. (3) The achievable climb gradient for the take-off condition (q_{To}) is greater than 0.027 and that for the landing condition (q_L) is greater than 0.024. (4) The take-off field length (D_{To}) is less than 1981m and the landing field length (D_L) is less than 1371m. The overall fuel balance coefficient is defined as the ratio of the fuel weight required for mission to that available for mission. The design variables are listed in Table 3.

TABLE 3. List of design variables

Design Variable / unit	Symbol	Lower limit	Upper limit
Wing area / m^2	S	111.48	232.26
Aspect ratio	AR	9.5	10.5
Design gross weight / 10^3kg	W_{dg}	63.504	113.400
Installed thrust / 10^3kg	T_i	12.587	24.948

Two disciplines, aerodynamics and weight, are considered in this problem. The dataflow between and in subsystems is analyzed in Fig. 14, where L/D_{To}, L/D_L, L/D_C are the lift-to-drag ratios for the take-off, landing and cruising conditions respectively, V_{br} is the cruise velocity with the longest range, R_{fr} is the fuel weight fraction required for mission, and C_{d0C} is the zero-lift drag coefficient for the cruising condition.

Two disciplines, aerodynamics and weight, are coupled. When the state variables in aerodynamics such as cruise velocity with the longest range, lift coefficients, zero-lift drag coefficients, skin-friction drag coefficients, lift-to-drag ratio are computed, some state variables in Weight such as R_{fr} should be known. Similarly, when the state variables in Weight such as useful load fraction, overall fuel balance coefficient, achievable climb gradient on take-off and landing, take-off field length and landing field length are computed, some state variables in Aerodynamics such as L/D_{To}, L/D_L, L/D_C and V_{br} should also be provided. In the Aerodynamics discipline, V_{br} is coupled with C_{d0C}. The dataflow between state variables and design variables can be seen in Fig. 15. Many details of equations in the aerodynamic discipline model and weight discipline model can refer to the reference (Zhang et al., 2008). The full description of them can be found in the references (Lewis & Mistree, 1995; Lewis, 1997).

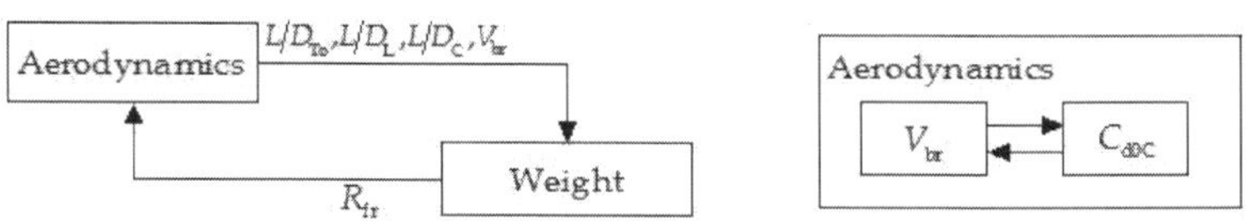

FIGURE 14. Dataflow between and in subspaces

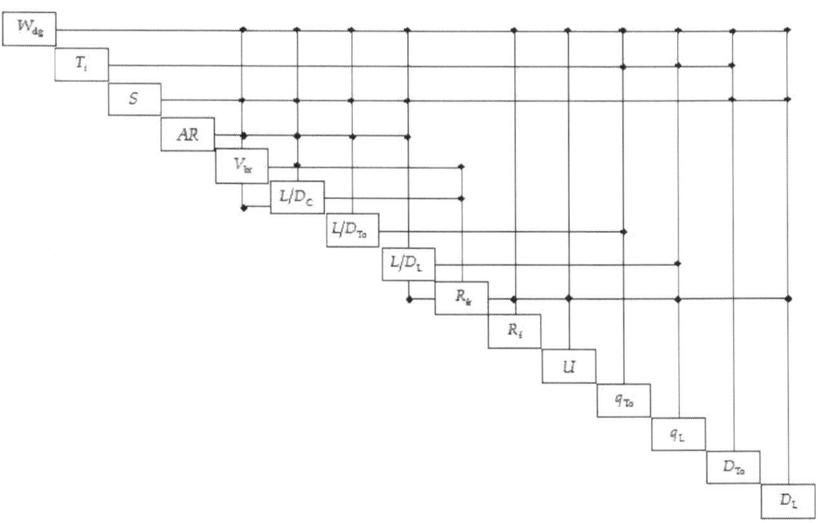

FIGURE 15. Dataflow between state variables and design variables

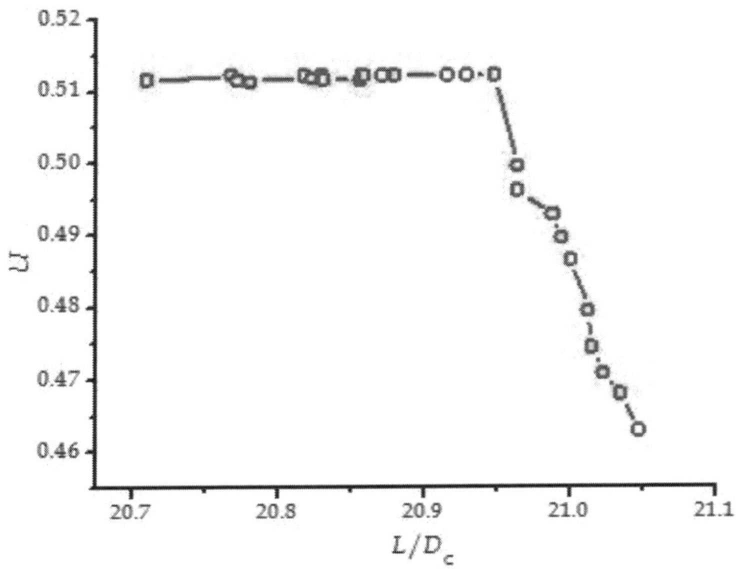

FIGURE 16. Pareto front obtained by using AWSCSSO

The Pareto front obtained using AWSCSSO is shown in Fig. 16. Each solution on Pareto front is obtained using CSSO with iterative subspace optimizations. Taking one of the optimal designs as example, the values of the design variables are: S=232.3m^2, AR=10.5, W_{dg}=113.4×10^{3}kg, T_i=16.75×10^{3}kg. The performance parameters of the aircraft in optimal design are as follows: C_{d0C}=0.01777, L/D_C=21.05, V_{br}=183.43m/s, q_{To}=0.03303, q_L=0.08804, D_{To}=1823m and D_L=1086m. Several conclusions can be made from these results. 1) The AWSCSSO method is primarily proved to be applicable for aircraft conceptual design. 2) The distribution of Pareto points is not so uniform as expected. These results are still very encouraging in general. The non-uniformity may be due to the additional constraint that changes the location to expected solution. Further study is still needed on how to achieve the balance between uniformity and convergence.

5. CONCLUSION

The CSSO method is one of the main bi-level MDO methods. Couples of CSSO methods for single- and multi-objective MDO problems are discussed in this chapter. It can be concluded that, (1) number of the system analysis can be greatly reduced by using the CSSO methods, which in turn improve the efficiency; (2) the CSSO methods enable concurrent design and optimization of different design groups, which can greatly improve efficiency; (3) the CSSO methods are effective and applicable in solving not only single-objective but also multi-objective MDO problems.

For the CSSO methods, although the RSCSSO method behaves more robust, it will actually reduce to a single-level surrogate modeling based MDO method since the subspace optimizations have little impact on the results. So the GSECSSO method is more promising as a bi-level method and worth further studying. The future study on the GSECSSO method will focus on improving its robustness and efficiency. For the multi-objective CSSO methods, the AWSCSSO method behaves better on obtaining widely-distributed Pareto points. The future work on the multi-objective CSSO methods will focus on improving the solution quality and also on testing it for more realistic engineering design problems.

REFERENCES

1. V. Aute, S. A. Azarm, 2006"Genetic Algorithms Based Approach for Multidisciplinary Multiobjective Collaborative Optimization," 11th AIAA/ISSMO Multidisciplinary Analysis and Optimization Conference, VirginiaAIAA 20066953

2. S. Azarm, W. C. Li, 1989". Multi-level, Optimization. Design, Global. using, Analysis,". A. S. M. E. Monotonicity, of. Journal, Mechanisms, in. Automation, Design, 1112259263

3. BolebaumC. L., 1991 "Formal and Heuristic System Decomposition in Structural Optimization," NASA-CR-4413,.

4. C. H. Huang, 2003 "Development, Multi. of-Objective, Subspace. Concurrent, Optimization, Methods. Visualization, Multidisciplinary. for, Ph. D. Design,", Dissertation, The State University of New York, New York,

5. C. H. Huang, C. L. Bloebaum, 2004". Incorporation, Preferences. of, Multi. in-Objective, Subspace. Concurrent, for. Optimization, Design,". Multidisciplinary, A. I. A. A. I. S. S. M. O. 10th, Analysis. Multidisciplinary, Conference. Optimization, York. New, AIAA 20044548

6. C. H. Huang, C. L. Bloebaum, 2007". Multi-Objective, Concurrent. Pareto, Optimization. Subspace, Multidisciplinary. for, A. I. A. A. Design,", Vol. Journal, 818941906

7. Kreisselmeier, G., Steinhauser, R.,1979 "Systematic Control Design by Optimizing a Vector Performance Index," IEAC

Symposium on Computer Aided Design of Control Systems, Zurich, Switzerland,.

8. I. Y. Kim, O. L. de Weck, 2004". Adaptive, Sum. Weighted, for. Method, Optimization,". Multiobjective, A. I. A. A. I. S. S. M. O. 10th, Analysis. Multidisciplinary, Conference. Optimization, York. New, AIAA 20044322

9. I. Y. Kim, O. L. de Weck, 2005". Adaptive, Method. Weighted-sum, Bi-objective. for, Pareto. Optimization, Generation,". Front, Structural, Optimization. Multidisciplinary, No, 149158

10. K. Lewis, F. ". Mistree, 1995Designing Top-level Aircraft Specifications-A Decision-based Approach to A Multiobjective, Highly Constrained Problem," 6th AIAA/USAF/NASA/ISSMO Symposium on Multidisciplinary Analysis and Optimization, Bellevue, WA, AIAA 1431

11. Lewis, K., 1997 "An Algorithm for Integrated Subsystem Embodiment and System Synthesis," Ph.D. Dissertation, Georgia Institute of Technology, Atlanta, Georgia, August

12. C. D. Mc Allister, T. W. Simpson, M. . Yukesh, 2000Goal Programming Applications in Multidisciplinary Design Optimization," 8th AIAA/USAF/NASA/ISSMO Symposium on Multidisciplinary Analysis and Optimization, CA, AIAA 20004717

13. C. D. Mc Allister, T. W. Simpson, K. Lewis, A. Messac, 2004". Robust, Optimization. Multiobjective, Collaborative. through, Optimization, Physical. Linear, 10th. A. I. A. A. I. S. S. M. O. Programming,", Analysis. Multidisciplinary, Conference. Optimization, York. New, AIAA 20044549

14. S. A. Orr, P. Hajela, 2005". Genetic, Based. Algorithm, Optimization. Collaborative, A. of, Configuration,". Tilrotor, A. I. A. A. A. S. M. E. A. S. C. E. A. H. S. A. S. C. 46th, Structural. Structures, Dynamics, Conference. Materials, Texas, AIAA 20052285

15. S. Parashar, C. L. Bloebaum, 2006". Multi-objective, Algorithm. Genetic, Subspace. Concurrent, . M. O. G. A. C. S. S. O. Optimization, Multidisciplinary. for, 47th. A. I. A. A. A. S. M. E. A. S. C. E. A. H. S. A. S. C. Design,", Structural. Structures, Dynamics, Conference. Materials, Island. Rhode, AIAA 20062047

16. Renaud, J.E. and Gabriele, G.A., 1993 "Second Order Based Multidisciplinary Design Optimization Algorithm Development," Advance in Design Automation,652347357

17. J. E. Renaud, G. A. ". Gabriele, 1993Improved coordination in non-hierarchic system optimization," AIAA Journal, 311223672373

18. J. E. Renaud, G. A. Gabriele, 1994". Approximation, non-hierarchic. in, optimization,". A. I. A. A. system, Journal, 321198205

19. R. S. Sellar, S. M. Batill, J. E. Renaud, 1996". Response, based. surface, subspace. concurrent, for. optimization, system. multidisciplinary, 34th. A. I. A. A. design,", Sciences. Aerospace, Meeting, AIAA 960714

20. Sobieszczanski-Sobieski, J., 1988 "Optimization by Decomposition: A Step from Hierarchic to Non-hierarchic Systems," Recent Advances in Multidisciplinary Analysis and Optimization, NASA CP-3031, Hampton,.

21. J. Sobieszczanski-Sobieski, 1990". Sensitivity, Complex. of, Coupled. Internally, A. I. A. A. Systems,", Vol. Journal, 1153160

22. R. V. Tappeta, J. E. Renaud, 1997". Multiobjective, Optimization,". A. S. M. E. Collaborative, of. Journal, Design. Mechanical, No, 403411

23. K. S. Zhang, Z. H. Han, W. J. Li, W. P. Song, . Bilevel, 2008Weighted. "Adaptive, Method. Sum, Multidisciplinary. for, Optimization,". A. I. A. A. Multi-Objective, Journal, 461026112622

Index